BY THE SAME AUTHOR

Seeing Is Forgetting the Name
of the Thing One Sees:
*A Life of Contemporary Artist
Robert Irwin*

The Passion of Poland:
*From Solidarity through
the State of War*

Shapinsky's Karma, Boggs's Bills,
And Other True-Life Tales

A Miracle, A Universe:
Settling Accounts with Torturers

Mr. Wilson's Cabinet of Wonder

Calamities of Exile:
Three Nonfiction Novellas

A Wanderer in the Perfect City:
Selected Passion Pieces

Boggs: *A Comedy of Values*

Robert Irwin:
Getty Garden

Vermeer in Bosnia:
Selected Writings

Everything That Rises:
A Book of Convergences

Seeing Is Forgetting the Name
of the Thing One Sees
(Expanded Edition):
*Over Thirty Years of Conversations
with Robert Irwin*

True to Life:
*Twenty-Five Years of Conversations
with David Hockney*

Uncanny Valley:
Adventures in the Narrative

Domestic Scenes:
The Art of Ramiro Gomez

WAVES PASSING IN THE NIGHT

Waves Passing *in the* Night

WALTER MURCH IN THE
LAND OF THE ASTROPHYSICISTS

Lawrence Weschler

B L O O M S B U R Y
NEW YORK · LONDON · OXFORD · NEW DELHI · SYDNEY

Bloomsbury USA
An imprint of Bloomsbury Publishing Plc

1385 Broadway   50 Bedford Square
New York   London
NY 10018   WC1B 3DP
USA   UK

www.bloomsbury.com

BLOOMSBURY and the Diana logo are trademarks of
Bloomsbury Publishing Plc

First published 2017

ISBN: HB: 978-1-63286-718-6
ePub: 978-1-63286-720-9

Library of Congress Cataloging-in-Publication Data is available.

2   4   6   8   10   9   7   5   3   1

Designed and typeset by Sara Stemen
Printed and bound in the U.S.A. by Berryville Graphics, Berryville, Virginia

To find out more about our authors and books visit www.bloomsbury.com.
Here you will find extracts, author interviews, details of forthcoming
events and the option to sign up for our newsletters.

Bloomsbury books may be purchased for business or promotional use.
For information on bulk purchases please contact Macmillan Corporate
and Premium Sales Department at specialmarkets@macmillan.com.

For Niwaeli
Our daughter's surprise midstlife twin,
newest member of the family

CONTENTS

Overture
ix

Walter Murch astride the Greenwich Meridian, 2016

OVERTURE

~~~~~~

FOR MORE THAN twenty years now, when he hasn't been doing anything else, Walter Murch has been tightening the bolts on his theory of gravitational astro-acoustics and refining the ever-more-accomplished PowerPoint across which he occasionally presents it before one august (albeit initially often quite dubious) audience or another. When doing anything else, more often than not he has simply been being Walter Murch, which is to say, at age seventy-three, arguably the most celebrated and admired film and sound editor in the world (nominated for nine Academy Awards and the winner of three; veteran of such classics as *THX 1138*, the *Godfather* films, *The Conversation*, *Apocalypse Now*, *The Unbearable Lightness of Being*, *The English Patient*, *Cold Mountain*, and, most recently, the Large Hadron Collider documentary *Particle Fever*). If he turns out to be right about that cockamamie theory of his, however, he may yet find himself in the running for a prize altogether grander than any Oscar.

But that's a big *if*—or rather, a big *double-if*. To begin with, whether he is in fact right about his theory, but even more immediately pressing, whether he can get anyone in the close-knit and largely closed-in community of professional astrophysicists even to give him, a rank outsider, and his unorthodox theory the time of day. Recently, however, there has been a slight if potentially momentous chink in the wholesale rejection of the kind of thinking Murch is advancing from within that very astrophysical community (of which more anon), such that the game at long last may actually be afoot.

PART ONE

Distant Music

MURCH LAUNCHES THE most recent iteration of his PowerPoint lecture (actually, these days he uses Keynote, though I myself have been hearing versions of this lecture for a good dozen years now, whenever I happened to be engaging Murch on a wide variety of other topics) by noting how there was a time, not that long ago, when the sort of thinking he is engaging in here (far from outlandish) was the very epitome of orthodox. From Pythagorean antiquity through the Middle Ages and well into the Renaissance, all learned gentlemen (and they were all mainly gentle*men*) were steeped in the fourfold classical curriculum known as the quadrivium, which is to say, arithmetic (pure number), geometry (number in space), music (number in time), and astronomy (number in space and time).

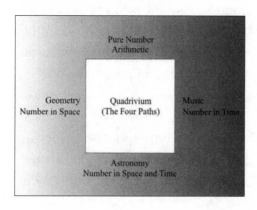

The quadrivium: the font of classical learning

From there, Murch goes on to project Galileo's iconic dashed drawing of 1610, documenting the latter's famous telescopic sighting of Jupiter with its four moons (all named after the Olympian god's human lovers: Io, Europa, Ganymede, and Callisto).

Galileo's drawing of Jupiter and its moons, 1610

Murch then jumps to a deep-space telescopic image, from 2008, of a star identified as HR8799 (129 light years away, sixty million years old) with three of its planets (labeled b, c, and d, orbiting 68, 38, and 24 astronomical units distant, respectively, from the star itself). (An astronomical unit, or AU, is the distance of the Earth from our sun, such that in the HR8799 system, the planet c is as far from its sun as Pluto is from ours: HR8799's is a very large system indeed, one of the main reasons we may even have been able to spot it at all.)

Noting how the actual circuits of orbiting bodies are almost always elliptical—as with Mercury, whose maximum distance from the Sun is 70 million km and minimum distance is 46 million km—Murch explains that he will be following the standard convention of citing only the average distance of the circling body across its entire orbit, the so-called "semi-major axis." Whereupon he projects a graphic representation of three concentric circles around the central star HR8799 at their correct relative distances, adding a fourth for a more recently discovered planet, e, the inmost one at 14.5 AUs, and then slides a similar representation, corrected for scale, of the relative orbits of those four moons around Jupiter (at 0.0028, 0.0045, 0.0072, and 0.0130 AUs respectively). And the four sets of concentric rings line up almost exactly, one atop the next—as do the relatively scaled orbits of Earth, Mars, and the asteroid Ceres (at 1.0, 1.52, and 2.77 AUs from the Sun respectively), when he thereupon slides in those.

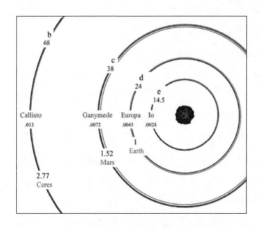

Comparative relative orbits of the satellites of the Sun, Jupiter, and HR-8799

An audible gasp invariably rises up from the audience right about here. At which point, Murch projects his next slide, the single word *APOPHENIA* in white against a black background. "The widespread tendency," he goes on to explain, "of human beings to see patterns where there are no patterns." Even Murch's cautionary provisos are elegantly framed.

WALTER SCOTT MURCH was born in New York City in 1943 and grew up there, in the Morningside Heights district surrounding Columbia University on the Upper West Side,

Walter Tandy Murch,
Carburetor, *1957*

the son of Katharine Scott, a secretary at the Ethical Culture School and subsequently Riverside Church, and Walter Tandy Murch, the eminent Canadian American painter.

His painter father's parents had been Toronto retail jewelers with a passion for music, and their three other sons became a conductor, a pianist, and a singer; but when the fourth, Walter Tandy, evinced a certain hesitation over the prospect of a lifelong career as a violinist, his mother suggested he might instead attend art school. "The boys were all going to be artists," as Walter Scott recently recalled in a monograph essay about his painter father, "and that was that." Walter Tandy took to painting with considerable gusto and moved down to New York City in 1927 to study at the Art Students League, remaining there owing to its more vibrant scene; with the passing decades, however, he arced away from that scene's ever-growing abstractionist tendencies. Following on from the urgings of his close friend (and fellow relative outlier) Joseph Cornell, Murch was by the late forties one of the few realists in gallerist Betty Parsons's stable (which otherwise included the likes of Jackson Pollock, Mark Rothko, Barnett Newman, Clyfford Still, and Ellsworth Kelly).

Walter Tandy Murch,
Sewing Machine, *1953*

Murch's realism, however, was of a highly distinctive sort (described at the time as "magic realism"). "Frequently the central focus of the painting," his son would later recall, "was a cast-off mechanical object from the past—a lock, a clock, a carburetor, an air filter—and he often stripped it of its casing skin, so that we were allowed to peer into its hidden skeletal structure. He once described the thrill of prying open an ancient door lock mechanism and discovering a moth's cocoon inside: 'I felt as if I was Howard Carter prying open King Tut's tomb,' he laughed." The resultant surrealistically tinged still lifes were conveyed across a subdued warm palette of reds and yellows and ochres, with an exceptionally narrow depth of field (perhaps owing to the fact that he was almost blind in one eye, due to a teenage

CLOCKWISE FROM LEFT:
Walter Tandy Murch, Katharine Murch,
Walter Murch at fourteen with his sister, Louise

football accident, such that the world through that eye was experienced as if "seen through translucent plastic").

"I don't paint the object anyway," the son recalls his father once telling him. "*I paint the air between my eye and the object.*" Elaborating, the son noted how

Turning air into paint—thinking of air as a transparent jellied matrix, molding itself around the objects, and then somehow transferring that insubstantial three-dimensional mold to the two-dimensional surface of the canvas—was my father's characteristically metaphysical way of coming to terms with (or nimbly sidestepping) the question: where does the meaning lie? Is it in the reality of the objects represented? Or is it the transubstantiated reality of this new object, the painting? Or somewhere else?

For his own part, in 1954, almost as soon as tape recorders first became commercially available to the general public, the painter's son, at around age eleven, became obsessed with *audio*-recording the world. As he recalls in the same essay, "I began recording random sounds in my local environment, at different speeds, then playing them backwards, upside down, back to front, and chopping the tape into bits and scotch-taping them back together in a different order. It was a kind of enthusiastic model-building—the sort of rabbit holes that twelve-year-olds fall into, and mostly soon pop out of." Then, one afternoon the following year, he turned on his radio and heard sounds that for a moment he thought were coming from his tape recorder, so similar were they to his own audio collages. The thrilling (to him) noises turned out to be an instance of the new-fangled *musique concrète*: a recording made in France by composers Pierre Schaeffer and Pierre Henry. "It was astonishing for me to find out that people—grownups!—in France were doing the same kinds of things I had discovered for myself accidentally, and it cemented in place a love of recording and mixing the sounds of the real world which has now persisted for sixty years." The private hermetic obsession of his twelve-year-

Cage

old self turned out to lay the foundation for his entire professional life.

A few years after that, his father took Walter to hear a lecture by John Cage, who would become another hero—Cage, the exacting polymorphous minimalist who would subsequently write:

> When I hear what we call music, it seems to me that some-one is talking. And talking about his feelings, or about his ideas of relationships. But when I hear traffic, the sound of traffic—here on Sixth Avenue, for instance—I don't have the feeling that anyone is talking. *I have the feeling that sound is acting.* And I love the activity of sound.

Sound acting across the surround, rebounding through the space between the object and the ear—that would become another of Walter's passions.

First, though, he headed off for college at Johns Hopkins, where he majored in nineteenth-century art history and Romance languages (French and Italian), graduating Phi Beta Kappa in 1965, after which he, along with classmates Matthew Robbins (the future director-screenwriter) and Caleb Deschanel (the future cinematographer), launched themselves out west (Walter cruising cross-country on his BMW R50 motorcycle with his new bride, a strikingly elegant and ribald young English nursing student named Muriel Ann—though everybody

*Newlyweds Walter and
Aggie, westward-bound,
1965*

called her "Aggie"—almost as tall as he was, "sometimes even taller," clinging to his back).

The boys were enrolled in the University of Southern California graduate film school, during a veritable golden age for film studies, where they joined up with the likes of John Milius (the truculent future screenwriter, parodied years later by way of the John Goodman character in *The Big Lebowski*) and George Lucas. At the same time, Francis Ford Coppola and Carroll Ballard were studying across town at University of California–Los Angeles (these in turn being the same years that Martin Scorsese was matriculating at that other American powerhouse, the film school at New York University). At one point, casting about for the subject for

The Zoetrope gang, San Francisco, 1969.
Coppola at top of ladder, Ballard mid-ladder, Milius at bottom
of ladder, Lucas behind them by window, Murch with pitchfork

a fifteen-minute student film, George Lucas asked Robbins and Murch if he could film one of their unproduced scripts, the story that became *THX 1138*. (Only many years after that did Murch come to realize that their dystopian tale dovetailed almost exactly with the plot and ethos of E. M. Forster's astonishingly prescient 1909 novella, *The Machine Stops*, which Murch insists they had not yet encountered at the time.)

After graduating from USC, Murch (along with Lucas) ventured up to San Francisco to join Coppola in founding the Zoetrope collaborative, a conspicuously un-Hollywood studio venture. (Indeed that's precisely why they'd all headed up to San Francisco, to get away from Hollywood.) Murch and Aggie presently settled in the decidedly laid-back hippie community of Bolinas, on the coast about an hour north of San Francisco, where they're based to this day. (Aggie, having raised four kids, run a horse barn and organic farm, and co-founded the local community radio station, KWMR, continues to produce prose, poetry, and radio pieces on the side.) Murch was involved in the sound edit and mix on Zoetrope's first feature, Coppola's *The Rain People*, and soon after undertook the sound design on another of Zoetrope's earliest ventures, George Lucas's first feature-length film, a low-budget expansion (cowritten with Murch) of his highly acclaimed student film, *THX 1138* (from the filming of which Lucas and he took an adventurous day off to participate in the Maysles Brothers filming of the Rolling Stones' horrific Altamont rock concert for their 1970 documentary *Gimme Shelter*, deploying a massive telephoto lens they'd only just been putting to use on *THX*).

While working on *THX*'s soundtrack, Murch stumbled upon one of the first of his many, many seminal editing discoveries, his so-called Rule of Two and a Half. Much of the film was shot on a white stage, across which grim black-suited robot policemen were meant to stride with particularly unsettling authority. But alas, the sound of the footfalls of the ordinary actors dressed up as robots, as recorded on the set, was anything but authoritative—it just sounded like people shuffling about. So Murch custom-built a pair of special metal shoes, fitted with springs and iron plates, and one night at 2:00 A.M., he wormed his way into the cavernously resonant halls of San Francisco's Natural History Museum, slipped on the iron contraptions, and proceeded to record a whole medley of his own varying footfalls. Returning to the editing bay, he then proceeded to meticulously lay the iron footfalls one at a time into the various scenes. "The whole process was insanely labor-intensive," Murch recalls, "like

Robots from THX-1138

embroidery," and he worried that he would never come to the end of it. Then, one day, he noticed how if you were dealing with one robot striding across the stage, you had to match the footfalls precisely, step by step, from one side of the screen to the other, in perfect sync. Likewise—and of course more exactingly yet—if there were two robots. If there were three or more, however, it didn't matter: you could just pile on footfalls in pell-mell profusion, and no one in the audience would be able to tell the difference. It seemed, he came to realize, that the human brain was capable of tracking two and a half streams of aural input at any given moment (two conversations, say, and a third half of a phone conversation), after which everything lapsed into an equivalence of noise. In subsequent years, he would come to learn that many early languages likewise had words for "one" and "two," but after that there would just be the word for "many." Similarly, the Chinese ideogram for "tree" looks like a single tree, the ideogram for "copse" is two trees, and "forest" is represented by three trees.

At any rate, *THX 1138*, when it was finally released, was especially celebrated for Murch's intricately modulated sound montage, and though today the film is widely recognized as George Lucas's most daring achievement as a pure cineaste, it basically bombed on its American release (perhaps owing to the pitch-dark, glaring-white bleakness of its imagined life-world, especially when contrasted with the, by comparison, chippier brightness of Stanley Kubrick's *2001: A Space Odyssey* of a few seasons earlier). Lucas learned his lesson, and for his next outing he tackled the decidedly lighter fare of *American Graffiti* (on which Murch also did sound). Meanwhile, however, *THX* proved a surprise *succès d'estime* at Cannes and garnered a year-long run at a Parisian movie theater, a circumstance that presently landed Murch in all sorts of hot water, arising from one scene in which the Robert Duvall character stumbles into what is basically a lab filled with fetuses burbling away in incubating jars. Murch had decided he wanted a repeating mechanical sound as part of the audio marinade into which he proposed to steep the scene, whereupon he recalled a recent suite of variations for a squeaking door and a woman's sigh by his old hero Pierre Henry, the avant-garde French *musique concrète* composer, from which he proceeded to borrow one squeak in particular, which he built into a repeating loop.

Henry

One day, Henry wandered into a showing of *THX 1138* at the Paris theater and, encountering the scene in question, bolted upright, exclaiming "*Sacré bleu!* That's *my* squeak!"—or words to that effect—and a few weeks later, Zoetrope received a letter from a formal Parisian barrister demanding satisfaction. Thus almost ended the burgeoning career of the young sound master Walter Murch. Somehow, wiser heads prevailed—it was determined that the squeak in question had been sufficiently modified to constitute fair use—but this is how it came to pass that (in addition to everything else) Murch is known for having provoked the first sampling lawsuit of the modern era, long before hip-hop and its myriad digital copyright imbroglios. Meanwhile, in large part because of the sheer extent of *THX*'s American flop, the entire Zoetrope enterprise, into which Coppola had poured the better part of his own personal fortune (such as it was then), now itself staggered toward bankruptcy, to avoid which Coppola was forced to take on a Hollywood studio project regarding which he had otherwise evinced no interest whatsoever: the film version of a recent bestselling mobster potboiler.

So from *THX*, Murch immediately shifted over to working on sound for Zoetrope's next film, the first of the *Godfather* movies, though, owing to union disputes between L.A. and San Francisco, his actual credit line on the film would be limited to the purposefully vague "post-production consultant." Nowadays, if you mention one of the film's most famous scenes—the one where Al Pacino as the Mafia family's golden boy Michael is sent to dinner at an Italian joint to meet with the corrupt police commissioner and the rival Mafia boss who owns him so as to exact vengeance on both of them for their near-assassination of his father—which culminates in Michael's momentarily withdrawing to the bathroom where a pistol has been hidden behind the toilet tank so that he can in turn come out firing, many fans will respond, "Oh yeah, that restaurant under the El tracks with the screeching trains going by!" The indelible El train effect, however, was neither in the book nor the script, nor was it considered during the shooting of the scene: it was entirely invented after the fact in postproduction by Walter Murch. "The thing was," as Murch explains, "Francis had wanted there to be no music at all through the entire scene, holding it in abeyance till Michael throws aside the gun and leaves the restaurant and gets into the getaway car, at which point the main musical theme was to come operatically surging to the fore. But the trouble was, much of the growing tension of the scene was taking place in Italian, which we were deliberately not subtitling, but which threatened to leave the audience at a bit of a loss, especially without music. So we decided to put in the sound of the El train. But to do that we had to work all the way back into the establishing shots, when Michael's car first arrives at the restaurant, and to retroactively create the effect of a train passing by unseen overhead: four successive waves of rumbling as the action proceeds through the whole dining room sequence, each one louder than the one before, with the terrible screeching which everyone remembers held in abeyance till the very climax of the scene, just before Michael fires his gun—though all of which, as I say, was only laid in many months after the scene had been shot."

Murch was becoming well known in the industry for his magisterial, almost Zen-like composure amid the roiling chaos of the editing process. ("Good editing," he likes to quote the director Victor Fleming, "makes a film look well directed; great editing makes a film look like it wasn't directed at all.") Though it turns out not to be true, as I have sometimes heard, that when Lucas now turned to his *Star Wars* franchise (on which Murch was unable to join him because that year he was working with Carroll Ballard on the screenplay of another upcoming Zoetrope project, the film of *The Black Stallion*), he based the character of Yoda on his old editing collaborator. "No, no," Murch demurs, laughing, when I raise the rumor with him, "that's absolutely not true, and I myself sure never heard it. But I did have a hand in inventing the name R2-D2." One night, he elaborates, during the laborious sound mixing of *American Graffiti,* as Lucas dozed in the corner of the screening room, Murch asked his assistant to rack up the second dialogue on the scene's second reel so they could look at the mix again.

Yoda interrogates R2-D2

What? asked the assistant. "You know," Murch said, "R2D2." At which point Lucas jolted awake, sputtering, "What did you just say!?"

Murch may not have been the model for Yoda, but there was a whole lot of Murch DNA in the makeup of the character (and even more so in the devotion to vocation) of Harry Caul, the wiretap sound protagonist portrayed by Gene Hackman in one of Coppola's next films, *The Conversation*— so much so that Murch, who managed sound design once again as well as, in this case, the picture edit, is often and not unfairly thought of as a virtual cocreator of the classic film, which gained him his first Oscar nomination (eclipsing recognition for his parallel sound mix work on the same year's *The Godfather: Part II*).

Murch continued on with Coppola across the epic work on *Apocalypse Now*, based on their Southern California film schoolmate John Milius's feverishly macho script, transplanting Joseph Conrad's *Heart of Darkness* way, way upriver into Vietnam. Indeed, so unapologetically war-pornographic was Milius's original script—Milius's own proud claim!— that Murch urged Coppola (as Coppola, for his part, is the first to confirm) to insert a My Lai–type incident to help complicate the film's overall moral tone, the famous scene where the navy boat comes upon a civilian sampan in the river, ferrying food to market, and when a young woman suddenly lurches for her pet puppy, the antsy American crew ends up shooting the whole place to bloody hell, a scene that Murch himself had a hand in writing. On the sound design side, perhaps the single most complex scene (arguably one of the most complex ever fashioned) proved

The Valkyrie scene from Apocalypse Now

to be the almost-ten-minute-long Valkyrie helicopter attack sequence, as Robert Duvall's Colonel Kilgore commands a fleet of air cavalry choppers into a swooping blitzkrieg assault on a coastal Viet Cong village, Wagner blaring terrifyingly from their speakers. Murch spent months separating out the cacophony of more than 175 separate soundtracks making up the sequence into six exactingly modulated premix streams (dialogue, helicopter, the Valkyries' music, small-arms fire, larger explosions, footsteps and other such minor clatter), but when he finally brought the six all together for the first rehearsal of the final mix, "everything seemed to collapse into a big ball of noise,"—jarring, inchoate, undecipherable—(talk about the imitative fallacy!). At which point Murch recalled his *THX* robo-footfall lesson: the Rule of Two and a Half. Go back and watch and, more to the point, listen to that scene when you next get a chance and you will notice how in the final version only two (or at most two and a half) of those six streams of sound rise to the fore at any given moment. If you are hearing the music and the thwack-thwack of the helicopters, there will be no dialogue; whereas if there is dialogue and small-arms fire, the helicopter thwack and the music will have been dialed way down, the whole thing registering as a crystal-clear evocation of chaos.

The Valkyrie theme was apparently contagious—or was it itself a symptom?—for somewhere in the seemingly endless process of filming his Vietnam *Gesamtkunstwerk*, Coppola got it into his head that, Wagnerlike, he, too, wanted to create his own Bayreuth, somewhere in the very middle of America, like Kansas, which would be the only place he'd permit his film to be shown, a vast purpose-built auditorium in which he could precision-control every facet of its presentation, in particular a revolutionary reconception of the theatrical film soundtrack format: the so-called six-track subsonic split-surround array which he and Murch spent months perfecting, with three speakers behind the screen (left, right, and center), two behind the audience (left and right), and a specially engineered sixth speaker for ultralow sounds (the body-penetrating thwack-thwack of the helicopters, say). In the end, Coppola was given to understand that no, that is not the way film distribution works, but the sound system survived (the only part of the dream that did), being retrofitted into all seventeen of the 70-mm theaters in which the film opened, and spreading out from there, presently to become the industry standard, now dubbed 5.1. Such that nowadays, when you go to a movie and you see the phrase "Dolby 5.1" in the closing credits, that's Walter Murch and the rest of the Coppola team. For his work on *Apocalypse Now*, Murch went on to win the Oscar for sound mixing as well as another nomination

Murch

for picture editing, and won the British Academy of Film and Television Award (BAFTA) in both categories.

I could go on: Murch endlessly, effortlessly tosses off such stories—see his widely acclaimed literary collaboration with Michael Ondaatje, *The Conversations.* Not only about Lucas and Coppola, but about the many other directors he has worked with: Fred Zinnemann on *Julia* (where his work was again recognized with both BAFTA and Oscar nominations); Phil Kaufman on *The Unbearable Lightness of Being*; Anthony Minghella (*The Talented Mr. Ripley*; *The English Patient*, for which, in unprecedented fashion, he received Oscars for both film editing and sound mixing; and *Cold Mountain*), Sam Mendes (*Jarhead*); among many others. But the thing about Murch is that that whole prodigious body of work (including a directorial outing of his own, the at-the-time underappreciated though increasingly well-regarded *Return to Oz* of 1985) is really only half the story—for, all the while, he is regularly engaged in all sorts of other pursuits off to the side.

His curiosity is phenomenally wide-ranging, the scope of his reading vast. He likes to carry around a pouch into which he stuffs a continuously renewed brace of paper slips containing Chinese fortune cookie–like pronouncements, from among which he regularly invites his acquaintances to reach in and sample. Over the years, here are some of the ones I myself have had the good fortune to spear:

The great battles are always waged where the maps overlap.
— NAPOLEON

Translate the invisible wind by the water it sculpts in passing.
— ROBERT BRESSON

Tell the truth as you see it; let beauty take care of itself.
— FRED ZINNEMANN

Truth never comes into the world but like a bastard,
to the ignominy of the one who brought her forth.
— JOHN MILTON

If at first the idea is not absurd, then there is no hope for it.
— ALBERT EINSTEIN

Music is the pleasure the human soul experiences from
counting without being aware that it is counting.
— GOTTFRIED LEIBNIZ

Seek simplicity, but distrust it.
— ALFRED NORTH WHITEHEAD

Science is the belief in the ignorance of experts.
— RICHARD FEYNMAN

Start every film with one certainty and a million questions;
end every film with a million certainties and one question.
— WALTER MURCH

Murch is the sort of person who has noticed (and will mention in passing) that five of the last seven presidents were born left-handed—a special interest of his because, as it happens, so was his father (or so anyway was he born, though, as was standard at the time, his parents tried to discourage the tendency by tying his left hand behind his back). Such parental interventions may have accounted not only for both the elder Murch's subsequent frustration at mastering the violin and the clumsiness with which he accidentally directed a passed football straight into his own eye, but also for the mild stammer that came to afflict him throughout his life, thereby further cleaving him from most of his macho abstract expressionist contemporaries. ("We now know that handedness and language share neuronal real estate," Murch fils explains, "and because handedness and speech turn out to develop simultaneously, the result is a frequent stammer if handedness is interfered with at this critical stage: a kind of permanent traffic jam of the neuronal pathways of speech.") He has figured out that blinks not only serve to moisten the eyes but also usually indicate a caesura between two thoughts in the mind of the blinker.

("You can see this in editing: the good actors blink between *their character's* thoughts, the bad ones between their own, as when, nailing a reading, the bad actor blinks wondering whether the director registered how good they just were; with good actors, the editing splice always occurs just before the blink.") He titled his slim volume on film editing—the Bible of most every film student, a sort of Zen and the Art of Filmmaking—*In the Blink of an Eye*, in part to highlight this fact but also in homage to that incredible scene in Chris Marker's *La Jetée* (Murch's knowledge of film history is equally prodigious). He knows (as we have seen) what *apophenia* means, and he also knows what *apoptosis* does (which is part of why, he once argued in a *Los Angeles Times* op-ed piece, that *tumor* would be a much better characterization for what is actually going on financially in what usually gets ever so mincingly characterized as a *bubble*), and that's only starting with the *a*'s.

He is the sort of person who, driving back and forth between his Bolinas homestead and various studios in San Francisco, will take to wondering, on a cubic-inch basis, which emits the most energy, the Sun or the human brain—and then will set to figuring it all out, looking up the requisite figures on Wikipedia and then calculating, *in his head*, that

WALTER MURCH'S SUNBRIGHT BRAIN CALCULATION

432,450	radius of sun in miles
5,280	feet in a mile
12	inches in a foot
2.74E+10	radius of sun in inches
2.06E+31	radius cubed
3.14	pi
1.33	= 4/3
8.61E+31	volume of Sun in cubic inches
.80	volume of human brain in cubic inches
1.08E+30	number of human brains in sphere the size of sun
20	output of human brain (20 watts)
3.84E+26	output of Sun in watts
2.15E+31	output of "brain sun" in watts
56,084	Difference ratio

This is how many more times the human brain is "hotter" than the same volume of the sun

indeed, if you reached in and pulled out an average 20-watt brain-size 80-cubic-inch handful from the inside of the Sun (determined, again, on an overall average basis), it would turn out that the human brain emits 56,000 times more wattage than the same volumetric amount of sun (or, anyway, as one wag noted when being so informed, *his* does).

"I was in Lyon at a hotel in 1986," Murch once recalled for me, when I asked him about his passion for Curzio Malaparte, the great, wildly unreliable magic realist Fascist, Communist, and possibly finally Catholic Italian midcentury journalist. "We were filming the Russian invasion scenes for *The Unbearable Lightness of Being* and I'd run out of things to read at night and so from a bookstore down the street I bought Hubert Reeves's book on cosmology, *L'heure de s'enivrer: l'univers a-t-il un sens?* And Reeves had used Malaparte's frozen horses story from early in *Kaputt* [Malaparte's fever dream of a memoir chronicling his travels along the Axis side of Europe's battlefronts during the Second World War] to illustrate the concept of *surfusion*, the ability of water, and other substances, to avoid phase shifts under certain circumstances: to go below the point of freezing while remaining still liquid, or above the point of boiling while likewise still remaining liquid. His point was that the universe itself is in a *surfused* state (more hydrogen and helium than there "should" be) because of the too-fast expansion after the Big Bang. And lucky for us, because without that much hydrogen, there would be no stars, at least not thirteen billion years after the Big Bang."

Just another typical moment in a conversation with Murch, in which the answer to one's question takes one so fascinatingly far afield that one almost forgets what one was asking in the first place. The Reeves reference, though, took him to the Malaparte book (where, in the third chapter, Malaparte records how a company of military horses on the Russian side of the Leningrad front stampeded out of their stables one frigid winter night when the stables caught fire after being hit by stray munitions, and, panic-stricken, the horses ran headlong into a lake on the very brink of freezing, where they then became trapped from one moment to the next when, without any intervening stages, the lake suddenly froze completely solid—or so Malaparte claims) which

ABOVE:
Gerri Davis, Ice Horses
(based on Malaparte)

RIGHT:
Malaparte emerging from a Finnish sauna

in turn got him, Murch, hooked on Malaparte such that in
the years since, alongside all his film and other work, Murch
has been seeking out rare untranslated bits of Malaparte and
translating the Italian's vivid, often almost hallucinogenic
prose, improbably (though to surprisingly bracing effect)
as poetry, as in this passage, taking place in the far north of
Finland, from the outset of "Partisans, 1944":

Around midnight,
we went to buy some cigarettes from the partisans.
And where the forest becomes thicker
about a mile upriver,
where big blocks of red granite
thrust upward through the grass,
we stopped, and waited.
It was raining.

The rain fell from the luminous midnight sky:
one of those transparent Arctic midnights of polished aluminum.
Muffled birdsong filtered through the branches
of red pine and white birch,
and the voice of the river rose and fell
like the light from a kerosene lantern.

Suddenly the partisans appeared:
young blond, tall, thin,
with red cheeks and blue eyes,
impeccably dressed in Allied uniforms:
jackets, overcoats, boots, and gloves
parachuted in from British planes.
We had brought bread, brandy,
reindeer milk and meat,
in exchange for cigarettes, soap and toothpaste.

And so forth. Murch recently published a collection of such translations, *The Bird That Swallowed Its Cage,* and he likes to talk about how similar the act of translating Malaparte into poetry seemed to him to his daytime job of film editing (the focus on pacing, the length and arc of the line, and the precise choice of where to break it). He's also spoken about the way he edits film standing up (like a brain surgeon, he says, or a conductor, or a short-order cook), whereas when he sets to thinking about or writing his own stuff, he does so lying down—but when Michael Ondaatje asked him about translating, he said he always did that sitting at a table.

Aside from illustrating yet another instance of Murch's protean activities, the thing that's striking about that whole Malaparte story, though, is who else, having run out of bedtime reading, visits a bookstore and settles on a volume asking *L'univers a-t-il un sens?* ("The universe, does it mean anything?"). The point is that Murch's astronomical and cosmological interests run way back, at least to 1986. Thus, for example, recalling a trip he once made to Rome and how he spent a long while one afternoon inside the Pantheon, the largest unreinforced concrete dome in the history of architecture, originally designed by the emperor Hadrian, who had been a sun-worshipping Mithraist (as I say, Murch just knows these sorts of things), Murch took to wondering about the impact such a sight might have had on the young Copernicus (b. 1473), who had been studying in Bologna (where he likely was exposed to the third-century B.C. heliocentric theories of Aristarchus) and from where he visited Rome in the Jubilee Year of 1500. What might the young ecclesiastical student have made of the light streaming down from the hole in the ceiling, and the concentric rows of scalloping welling out from that hole? Copernicus only allowed his manuscript *De revolutionibus orbium coelestium* (*On the Revolutions of the Heavenly Spheres*) to be published in 1543, toward the very end of his life—and

The Pantheon in Rome

incidentally, that's where our current usage of the word *revolution* comes from, Murch is likely to interrupt himself at this point to explain, from the way Copernicus's manuscript describing the circular orbits, the revolutions, of the planets around the Sun upended the established order of things: hence, by way of analogy, any similar subsequent upending of the order of things hearkens back to the *title* of that book and not, as is often assumed, to any supposed 360-degree rotation of the established order, which, if you think about it, wouldn't make sense in any case (except maybe to The Who). Anyway,

in the tenth chapter of that manuscript, Copernicus's text accompanying the first diagrammatic illustration of his theory of planets orbiting concentrically around the Sun describes how "At rest in the middle of everything is the Sun," and goes on to ask, "For in this most beautiful temple [*in hoc pulcherimo templo*], who could place this lamp in another or better position than the center, from which it can light up the whole at the same

Copernicus

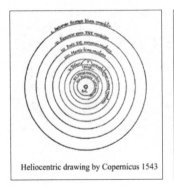
Heliocentric drawing by Copernicus 1543

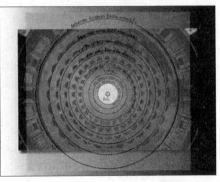

Copernicus's diagram of the solar system overlaid on the Pantheon

time?" Particularly struck by the use of the word *lamp* in that passage, since he knew that the open oculus at the top of the Pantheon was referred to at the time as its "lantern," Murch now, operating on a hunch, managed to slide a photograph of that dome, seen from inside, over Copernicus's diagram, and the circles lined up almost perfectly. Apophenia? Perhaps. Who knows? But as Murch subsequently said, in typical understatement, "Seeing that alignment was one of those wonderful moments where you suddenly feel a strong current of connection with the past."

Though it wasn't so much Copernicus whom Murch came to place among his greatest heroes as Johannes Kepler, whom he first really got hooked on, a few years after encountering that Reeves book, by way of Tim Ferriss's *Coming of Age in the Milky Way*, thereafter hoovering up pretty much anything he could find on the seventeenth-century astronomer (fodder, he'd tell himself, for a possible film treatment someday).

A few years ago, Murch penned a short historical essay of his own entitled "Oblivion or Immortality," which begins, "The two men in the coach were both 28 years old, born within a few months of each other in 1571. Frederik was Danish and Johannes was German, and for different reasons they now found themselves jostled together, in early June of 1600, traveling from Prague to Vienna." The carriage mates on this three-day trek turn out to be, yes, Johannes Kepler, and a Danish wastrel named—wait for it—Frederik Rosenkrantz. Both had just been to visit the great Danish astronomer Tycho Brahe, himself recently exiled to the court of the Hapsburg emperor Rudolf II, the young Kepler because he aspired to work with the foremost astronomical observer of his time (and to get the chance to survey his unparalleled observational logs) and Rosenkrantz because he happened to be Brahe's third cousin! Murch goes on to rehearse the respective fates of the two young gentlemen, expanding at some length on Kepler's myriad eventual insights and achievements and on his particularly beguiling cast of mind, reflected in the sweet savor of his prolific prose, as in this passage from his 1606 *De Stella Nova*:

> Yesterday, when weary with writing, I was called to supper, and a salad I had asked for was set before me by my wife. "It seems, then," I said, "if pewter dishes, leaves of lettuce, grains of salt, drops of water, vinegar, oil and slices of eggs had been flying about in the air from all eternity, it might at last happen by chance that a salad would finally appear." "Yes," responded my lovely, "but not so nice as this one of mine."

For his part, Rosenkrantz—who had been cast out of Denmark a few months earlier on account of his role in

RIGHT:
*Tycho Brahe with
family crests*

BELOW:
Kepler

an egregious sex scandal and only spared the stripping of his nobility (along with two fingers) on the promise that he would instead go join the Austrian military campaign against the Turks, who were presently preparing to lay siege to Vienna—well, as Murch notes:

> Far from having a salad set before him by a lovely wife, Frederik was killed in Moravia two years later, in 1602, attempting to separate two dueling comrades-in-arms:

for him, oblivion seemed inevitable. But in London that same year 1602, a new play was being performed about a Danish prince. The playwright had needed names for two characters who were plotting the prince's death, and he recalled a high-profile diplomatic mission from the King of Denmark to the Queen of England ten years earlier, in 1592. Two young courtiers on that mission, both relatives of Tycho Brahe, had made spectacles of themselves in the taverns and fleshpots of London, and the unintended consequence of their carousing was immortality, though of a dubious kind: twenty-one-year-old Frederik Rosenkrantz and his friend Knud Gyldenstierne had somehow attracted the attention of William Shakespeare and, in the play *Hamlet*, yielded up their surnames to posterity as bywords for bantering fecklessness and sycophantic treachery.

In 1994, while working in London on Jerry Zucker's film about Lancelot and Guinevere, *First Knight*, though still in thrall, over to the side, to his Keplerian passion, Murch found his way to Arthur Koestler's polemically revisionist masterpiece *The Sleepwalkers* (in which it is indeed Kepler,

LEFT:
Shakespeare

rather than Copernicus or Brahe or Galileo, who stands out as the true hero of the story). Deep in the back notes section of that book, Murch came upon a glancing reference to another subsequent pair of German Johanns, Johann Daniel Dietz (1729–96) and Johann Elert Bode (1747–1826), and a curious little theory of theirs, the fate of which was soon to *really* start obsessing him—which brings us, by an admittedly wide albeit hopefully commodious vicus of recirculation, back to the very next slide in that lecture of his.

FOR NOW, AS Murch advances his digital carousel, up pops a new image: that of a dapper Enlightenment gentleman straight out of central casting—as Murch explains how, one day in 1766, Johannes Dietz (known forever after as Titius, following the custom among academics of his time of Latinizing their names after they had received their doctorates) published a German translation of Charles Bonnet's *Contemplation de la Nature,* into which he had interleafed (as was also a custom at the time) some of his own bits of speculation, and in particular a remarkably simple proposal that initially read, as Murch says, "like a cookbook recipe rather than an assault upon the hidden structure of the universe":

Titius

Start with doubling series	0	1	2	4	8	16	32
Multiply by three	0	3	6	12	24	48	96
Add four	4	7	10	16	28	52	100
Divide by ten	0.4	0.7	1	1.6	2.8	5.2	10
Relative distances of planets	0.39	0.72	1	1.52		5.2	9.6

Mercury *Venus* *Earth* *Mars* ??? *Jupiter* *Saturn*

Titius's Rule, 1766

Start, that is, with the doubling series (0, 1, 2, 4, 8, 16, etc.), multiply each number by three, add four, and divide by ten. The resulting sequence of numbers, Titius went on to note, bore a striking resemblance to the average relative distances of the planets from the Sun, when compared to the Earth's distance, which is set as 1 for the sake of convenience (this "1" being known in astronomy, as we have seen, as the astronomical unit, abbreviated as AU). So: Mercury's actual average distance from the Sun is 0.39 AU; Venus's 0.72; Earth's 1.0 (by definition of AU); Mars's 1.52; then at 2.8 an empty space; Jupiter's 5.2, and Saturn's 9.54. ("Saturn," as Murch notes, "being the furthest out of the 'naked-eye' planets, which were all known to humans from before the invention of writing, perhaps even before the invention of language, such that, as late as 1766, we still knew of no other planets than these.") Asking, "Would the Lord have left a space empty?," Titius went on to predict that there had to be something lurking there at 2.8, between the orbits of Mars and Jupiter, as yet undiscovered.

Planet	x	AU	Actual	Accuracy	Deviation
Mercury	-∞	0.400	0.387	96.75%	3.2%
Venus	0	0.700	0.723	103.29%	3.3%
Earth	1	1.000	1.000	100.00%	0.0%
Mars	2	1.600	1.524	95.25%	4.8%
Empty	3	2.800			
Jupiter	4	5.200	5.203	100.06%	0.1%
Saturn	5	10.000	9.537	95.37%	4.6%
	6	19.600			
	7	38.800			
				98.45%	2.66%
				Collective	Individual

Solar System, 1766

Titius was merely a professor of physics and biology at Wittenburg University, but six years later, Johann Bode, the Neil deGrasse Tyson of his time, the head of the Berlin Observatory and a great proselytizer of astronomy, took up Titius's speculations, addressed an obvious problem (while all the other numbers in the initial sequence are doubles of the previous one, two times the first number in the sequence, 0 doubled, is obviously not 1 but rather 0), and expressed the procedure in a more strictly conventional formula, which he modestly claimed as his own, Bode's law, to wit:

$$\frac{4 + (3 \times 2^n)}{10}$$

The *n* in Bode's version, he explained, was meant to stand for the numerical progression:

$$-\infty, 0, 1, 2, 3, 4, \text{and so forth}$$

and when those numbers were plugged into the new formula, the resulting sequence was indeed the same as Titius's: Mercury's 0.4 turning out to be at the limit of negative infinity for this formula, and Venus's 0.7 resulting from n = 0, Earth's 1.0 from n = 1, etc. There was still a peculiarly arbitrary jump, however, this time all the way from $-\infty$ to 0 in one fell swoop. About which, more later.

Bode had also predicted that something would soon show up at 2.8, but instead, in 1781, the German British musician and genius telescope maker William Herschel, out stargazing with one of his self-built telescopes (the best

in the world at the time), stumbled upon an altogether different celestial entity, a new planet (the first ever to be discovered since humans began gazing at and pondering the nighttime sky!) whose distance he was eventually able to fix at 19.2 AUs, within 2 percent of 19.6, the next number (after Saturn) in the Titius-Bode sequence, obtained when n = 6. Herschel had

Bode

Herschel

called his new planet George (hoping to extract patronage from the British king George III, and indeed British astronomy books would still be calling it Georgium Sidus, or "George's Star," until well into the middle of the nineteenth century), but Herschel's birth-countryman Bode instead convinced him to follow convention and ("to the eternal giggling delight of seventh grade boys everywhere," as Murch acknowledges) name it Uranus, the name deriving from that of Ouranos, the Greek god of the sky and mythical father of Saturn, who in turn was the father of Jupiter, who in turn was the father (moving inward) of Mars, Venus, and Mercury.

As for vindication of their initial prediction of something at 2.8, Titius didn't live to see it (he died in 1796), but

Piazzi

Bode did, for in 1801, the Sicilian priest-astronomer Giuseppe Piazzi, professor of astronomy at the University of Palermo, accidentally spotted a planet at 2.77 AUs, only 1 percent off the Titius-predicted 2.8 AUs. He named it Ceres, in honor of the ancient goddess of agriculture and protector of Sicily, and it turned out to be the largest in what would subsequently come to be

understood as a belt of asteroids. From that point forward,
for the next several decades, having advanced a theory that
had proven both descriptive and predictive ("the gold standard in science," Murch notes), Bode got to bask in the conviction that (give or take a measly Titius), he had indeed
authored an actual law.

Planet	x	AU	Actual	Accuracy	Deviation
Mercury	-∞	0.400	0.387	96.75%	3.2%
Venus	0	0.700	0.723	103.29%	3.3%
Earth	1	1.000	1.000	100.00%	0.0%
Mars	2	1.600	1.524	95.25%	4.8%
Ceres	3	2.800	2.766	98.79%	1.2%
Jupiter	4	5.200	5.203	100.06%	0.1%
Saturn	5	10.000	9.537	95.37%	4.6%
Uranus	6	19.600	19.191	97.91%	2.1%
	7	38.800			
				98.43%	2.41%
				Collective	Individual

Solar System, 1768

But, not so fast, objected Carl Friederich Gauss, the
twenty-three-year-old mathematical prodigy (possessed of
a genius both so precocious and sublime that he is often
referred to as the Mozart of Mathematics). Gauss is perhaps
most famous among the mathematical laity for his prodigious
feat at age ten, upon being asked along with his classmates
by their teacher to calculate the sum total of all the integers between 1 and 100, of almost immediately, and without
any evident scribbled calculations, announcing 5,050; when
asked by his flabbergasted teacher how he had come up
with the correct sum so quickly, wee Carl Friederich replied:
"Easy, 100 plus 1 is 101, 99 plus 2 is 101, 98 plus 3 is 101; 50

Gauss

times 101 is 5,050." Just over a dozen years later, Gauss in fact played an important role in confirming the existence of Ceres. Piazzi had been able to follow the new object for only three degrees of the arc of its orbit before it slipped behind the Sun, and who knew when or where it was ever going to reemerge. Gauss, though, applied himself to the problem, generating an entire branch of mathematics in the process (published a few years later as his theory of celestial movement, the cornerstone of astronomical computation to this day), whereupon he predicted precisely where and when, as indeed proved precisely correct.

Nevertheless, Gauss was having none of Titius-Bode.

Gauss's objections basically amounted to three. (And I should perhaps point out here that in all the foregoing and in what follows, I am relying not solely on this most recent iteration of Murch's lecture but also on several earlier versions, as well as on an unpublished manuscript Murch drafted in 2004 and whose title, "Skybound," he derived from Kepler's 1630 epitaph, "Skybound was my mind.")

To begin with, as Murch relates, Gauss objected to the entirely arbitrary and unwarranted leap in Bode's sequence from negative infinity to 0, which actually made no more mathematical sense than Titius's original doubling leap from 0 to 1.

Gauss's second objection was essentially that Bode's formula was just too ungainly—nature doesn't indulge in all those arbitrary constants (2, 3, 4, 10); Newton's formula for universal gravitation, for instance, had just one constant, yet it applied to an infinite number of situations. Continuing, Gauss observed that, furthermore, Titius and Bode were dealing with just too few "planets" (eight) from which to derive a reliable law—any chance arrangement of a small number of objects (coins tossed on a table, birds singing on a tree branch), he argued, could be accounted for if you allowed yourself, as Bode and Titius had, a sufficient number of arbitrary constants.

Finally, there didn't appear to be any physical explanation for why this clunky formula might be playing out in the actual physical world.

Gauss concluded forcefully that absent adequate responses to those three objections, Titius-Bode could only be considered numerology. Nevertheless, widespread belief in the efficacy of the law persisted for several decades more.

As the years passed, however, astronomers noticed that something seemed to be tugging on this new planet Uranus, periodically forcing it to speed ahead or lag behind where it was supposed to be in its orbit by minimal and yet detectable degrees, and the existence of a planet yet farther out was hypothesized. Two astronomers, one in England and the other in France, unaware of each other, closed in on its location, triangulating from the perturbations in Uranus's orbit and assuming a Titius-Bode distance of 38.8 AUs, from

Planet	x	AU	Actual	Accuracy	Deviation
Mercury	-∞	0.400	0.387	96.75%	3.2%
Venus	0	0.700	0.723	103.29%	3.3%
Earth	1	1.000	1.000	100.00%	0.0%
Mars	2	1.600	1.524	95.25%	4.8%
Ceres	3	2.800	2.766	98.79%	1.2%
Jupiter	4	5.200	5.203	100.06%	0.1%
Saturn	5	10.000	9.537	95.37%	4.6%
Uranus	6	19.600	19.191	97.91%	2.1%
Neptune	7	38.800	30.050	77.45%	22.6%

Solar System, 1846, with Neptune

which they in turn calculated the phantom planet's mass and the ellipticity of its orbit. In August 1846, checking up on predictions of the Frenchman Urbain Le Verrier, astronomers at the late great Bode's own observatory in Berlin sighted the planet that would in turn get named Neptune (another scion of Saturn, along with siblings Jupiter, Pluto, and Ceres). Yet it quickly became apparent that something was wrong, for the planet was not nearly as massive as had been predicted; nor was its orbit as elliptical (astonishingly the two miscalculations had cancelled each other out, allowing for the planet's initial sighting). Its actual distance from the Sun was recalibrated at 30.05 AUs, almost 25 percent less than Bode's predicted figure. And from one moment to the next, Bode's "law" lay in ruins.

"So much so," Murch points out, "that almost immediately, exactly eighty years after its first promulgation by Titius, the theory took on the reputation of a complete folly, a fever dream, exactly as Gauss had insisted: pure

Planet	x	AU	Actual	Accuracy	Deviation
Mercury	-∞	0.400	0.387	96.75%	3.2%
Venus	0	0.700	0.723	103.29%	3.3%
Earth	1	1.000	1.000	100.00%	0.0%
Mars	2	1.600	1.524	95.25%	4.8%
Ceres	3	2.800	2.766	98.79%	1.2%
Jupiter	4	5.200	5.203	100.06%	0.1%
Saturn	5	10.000	9.537	95.37%	4.6%
Uranus	6	19.600	19.191	97.91%	2.1%
Pluto	7	38.800	39.529	101.88%	1.9%

Solar System, 1930, without Neptune

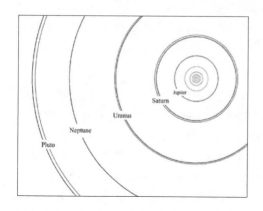

The outer planets of the solar system, Bode-predicted and actual orbits

numerology. And it has remained so to this day. Students in Astronomy 101 classes nowadays regularly get warned against an entire tenor of speculation, 'Don't think like that or you'll end up like Titius-Bode,' they're told. The whole thing has scientific cooties.

"There have been regular attempts to resuscitate the theory," Murch says, "and professional astrophysicists will not even entertain them. Reputable journals have policies against accepting papers on the subject: untrue: cannot be happening. So much so that in 1930, a bit more than eighty years after the discovery of Neptune, when Pluto was discovered by Clyde Tombaugh, a young astronomer in Arizona, nobody even bothered to note that it came in at 39.53 AUs from the Sun, less than 2 percent off the original Titius-Bode prediction for Neptune of 38.8."

NOT BEING A trained astrophysicist and perhaps therefore never having contracted the allergy, Murch thought he saw a way clear from the moment he encountered reference to Titius-Bode in that footnote in Koestler's *Sleepwalkers*. As he wrote in his journal that very first night in September 1994:

9/12/1994
Bode's Law: The law of planetary intervals, discovered in the late 18th century • $1 \times 2^n \times 3 + 4$ • Why is this so, and is it true for other planetary systems? • Does it have anything to do with Kepler's "divine solids" which he felt (originally) to be the determining factor in establishing the distance between planets • Kepler's solids were a classic example of a false bisociation yielding greater truth. (like Columbus and his false—too low—idea of the circumference of the earth) • If Bode's law is reformulated as (2^n), does it work? If not, how far off is it? With negative as well as positive values for n.

Now, in his lecture, Murch acknowledged that any way forward with Titius-Bode would need to come to terms with Gauss's completely valid objections. To begin with, though, regarding the leap from $-\infty$ to 0, Murch suggested just getting rid of it: simply acknowledge that there could indeed be an infinite number of *unoccupied* orbits between negative infinity and zero (and, for that matter, infinite orbits going in the other direction as well toward positive infinity). This ran contrary to the *horror vacui* of both Titius and Bode, traditional still religiously flecked eighteenth-century scientists who were both of them steeped in a principle of parsimony whereby God would never have created orbits that he did not fill. That had been the very basis for Titius's conviction that something was going to show up between Mars and Jupiter at 2.8 AUs. Murch offered an analogy with the subatomic realm, as it came to be understood in the early twentieth century, where there are all sorts of potential orbits around atomic nuclei that might or might not be filled by electrons at any given moment. "We thus simply remove the arbitrary jump from $-\infty$ to 0 in the original formula," he concluded, "and replace it with a true infinite series. This makes the formula more mathematically respectable, but only by proposing that there can be an infinite series of ever-smaller unoccupied (virtual) orbits between Venus and Mercury. We will only discover whether this is a valid assumption if and when we apply Titius-Bode to other systems and find that some of those orbits indeed turn out to be occupied."

From there, Murch turned to the clunkiness of Bode's formula with its suspicious surfeit of constants (2, 3, 4, and 10). Wasn't the problem, Murch now asked, caused by the nature of both Titius's and Bode's conceptions, the way that Earth's distance was the standard of comparison? The arbitrary insistence on Earthly importance implied in the deployment of 1 AU as the anchoring unit of measure in turn skewing the resultant formula. "The original Bode formula had not only to generate a series of numbers," Murch observed, "it also was forced to compare those numbers to a value (Earth's distance) in the middle of the series itself."

If instead of basing the entire system on 1 being the arbitrary AU distance (with Mercury as 0.4 and Venus as 0.7), what would happen if one cast the inward limit of the sequence as "1," with the distance from the Sun to that point, which Murch proposed calling beta (β), becoming the standard of measurement. How many βs would equal a single AU (how many 0.4s in 1.0?). The answer was simple: 2.5. Multiply Bode's entire formula by 2.5 and all sorts of things cancel out and you get something much simpler and more conventional-looking, to wit:

$$1 + 2^{n-2} \times 3$$

which, since Murch has established that n runs the full gamut from negative to positive infinity (and following the mathematical rule that if you have an infinite series where n is defined as all integers from $-\infty$ to $+\infty$, then writing 2^{n-2} gives you the identical series of numbers as writing 2^n), can be written out even simpler as:

$$1 + 2^n \times 3$$

A perfectly respectable formula of the sort that Newton or Gauss might easily have been able to live with, and one that nevertheless yields exactly the same sequence of ratios between the resultant orbits as did that original Bode formula.

Murch suggests that another way of thinking about β, which is to say in this instance the distance from the sun to the limit of negative infinity in the revised Bode-Murch formula, is to hypothesize that any object falling inward from that border would over time have gotten sucked into the Sun itself by the Sun's own gravity. In that sense, β can be thought of as the minimum distance that an object would have to be from the Sun (or any other gravity-exerting object) to have a long-term stable orbit and not eventually get dragged into the center (or else, as in the so-called Roche limit, get pulled in so fatally tight that tidal forces stretch the object to smithereens, scattering the results in fine dust rings, as with Saturn's)—a kind of watershed, in effect. There may well have once been other protoplanets closer in than Mercury, but they all slid into the Sun, and it's no surprise, then, that Mercury finds itself quite close to the β distance: it was the first (or last, depending on how you look at it) such protoplanet left standing. (Not that there always has to be a planet filling that conceptual orbit—and as we shall see, in most systems there isn't—though it happens that Mercury does.)

From that β point outward, Murch goes on, there occur a series of orbit doublings, at first across a series of infinitesimally small potential orbits (not necessarily filled) until one gets to orbital shells at, say, $n = -9$, which is twice the distance (as measured from β) as $n = -10$; $n = -8$ is twice the distance from β as $n = -9$; and so forth, until one arrives at $n = -2$ (Venus's orbit) and then $n = -1$ (Earth's orbit), and n = 0 (Mars's orbit) and so forth. Indeed, Murch points out, this is exactly the sort of thing we see in the rings of Saturn, millions of baseball-size objects compressed into compoundingly narrower and narrower corrugated orbits until they reach the point where Saturn's own gravity (swallowing up anything closer in) has cleared out an empty cordon sanitaire:

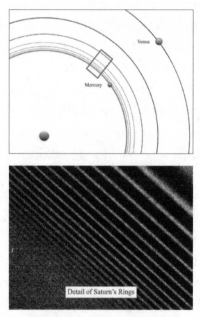

Tightening potential orbits
nearest the Sun remind Murch
of the inner rings of Saturn

"In effect," Murch goes on to suggest, "what we are seeing is a doubling series of waves, a series of swells and troughs—a statistical corrugation certainly, if not a physical one, and definitely not a planar one. If anything, it is spherical, like the layers of an onion, but a sphericity that can be schematized in a plane as if the system were behaving like this

Compounding notional waves welling out from a central object

with the orbiting objects generally tending to fall in the troughs between the peaks of the waves. Phrased differently, there appear to emerge, over long stretches of time, hundreds of millions of years or more, "distances where planets tend to find stable orbits and distances where they don't, places that are fertile for orbiting objects, and other places, the peaks of the waves, as it were, that are not, that are barren." The formula for the *troughs* of the waves is the one we have been looking at $(1 + 2^n \times 3)$, "and now we can see more clearly the reasons for the numbers in the formula—what they are accomplishing: The number 1 puts the radius β at a certain distance from the center of the Sun (or whatever is the central body), the 2 raised to various powers gives us the peaks of the doubling waves starting from that point β, and the 3 puts us at the midpoint between the peaks of those waves."

But, Murch goes on to point out, the formula for the *peaks* of the waves, the area of barrenness, is simpler still:

$$1 + 2^n$$

At which point he projects a visualization of the state of play thus far:

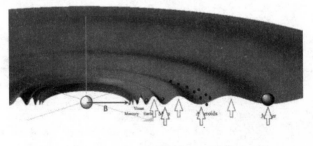

$$\beta(1+2^n) \quad \beta(1+2^n \cdot 3)$$

The state of play thus far

The thing is, though, and here is where Murch's status as one of the foremost sound designers of our time comes into play (whether or not he has any standing as an astrophysicist), the $(1 + 2^n \times 3)$ formula has a history; it contains within it Pythagoras's original reckoning for the musical octave, where any key ♪ doubles frequency with each octave: $♪(2^n)$. Thus, for example, if A is 440 hertz, then A (2^n) plays out as 110 when n = –2; 220 when n = –1; 440 when n = 0; and 880 when n = 1.

The key point here being (and this is acoustician Murch's seminal original insight at this point in his argument) that the ratio between the distances of the Bodean orbits (one radial distance in kilometers divided by the other) turns out to be exactly the same as the ratio between the musical notes (which is to say, one frequency in cycles per second divided by the other), such that, if you apply Murch's formula β $(1 + 2^n \times 3)$ you can lay such musical notes onto the solar system, which Murch proceeds to do, generating a notional music of the spheres, indeed showing how the various planetary orbits can be shown to align with a series of harmonically pleasing tones, with the solar system as a whole constituting a single wide chord, indeed one based on what is known in musical circles as the seventh chord. To wit, if we arbitrarily consider Mercury's orbit to be at the note C, then Venus is at the B-flat above that, and Earth the E above that, and Mars at C two octaves above Mercury's, the asteroids at B-flat two octaves above Venus's, Jupiter at A-flat above that, Saturn at A-flat one octave above Jupiter, Uranus at G, and then Pluto at G above Uranus... All of which, at other moments, Murch loves demonstrating on a piano, whose eighty-eight keys, conveniently, can just contain the whole spectrum from Mercury to Pluto.

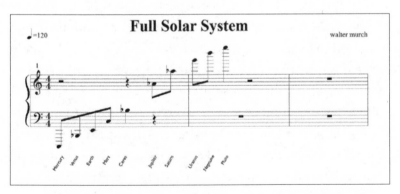

The Solar System, reflecting actual ratios across the entire keyboard

Or, a more "playable" version, with the planets transposed into the mid-keyboard, with the ratios between the first five and the last five being correct in sequence, but a transposition occurring between Ceres and Jupiter, where

Jupiter is in fact three octaves lower than it would be if the correct ratio were maintained between it and Ceres:

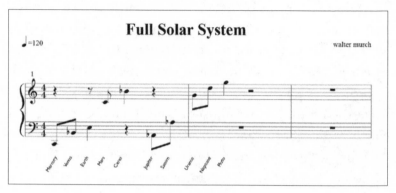

The Solar System, recalibrated mid-keyboard (for easier playing)

At this point, Murch will again insist this is *notional* music, a statistical music, not one actually playing out there in the vacuum of space. It is not, for example, the Music of the Spheres as put forth by Pythagoras and medieval astronomers who imagined that the Earthly sphere (made up as it was by four elements: earth, air, water, and fire) was nested centrally inside a sequence of perfectly pure crystal spheres (spheres made out of an otherworldly fifth essence, the quintessence, hence the origin of *that* word) into each of which was embedded, like a jewel, the Moon and the various "naked eye" planets, including the Sun—one such globe nested inside the next, all of them moving at different rates in relation to one another and carrying their embedded planets with them. And it was the sound of the various spheres rubbing against each other as they turned, one in relation to the other—the Moon's sphere revolving the fastest and Saturn's the slowest—that constituted the divine

harmonium, the sublimely gorgeous Music of the Spheres
(for how could anything fashioned by God be anything less
than sublimely beautiful?).

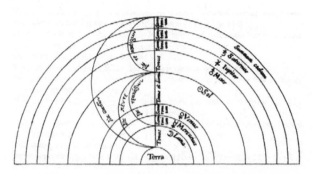

The Music of the Spheres, according to Pythagoras

Murch's music of the spheres was *not* this, he reiterated, but
it was, he suggested, surely *something*. Of which more anon.

Murch thereupon turned to the rest of Gauss's second
objection, that there simply weren't enough instances upon
which to base any sort of coherent theory. The fact was, as
Murch pointed out, in the years since Bode fell into such
disrepute, astronomers have in fact been able to establish
precise figures for the orbits of all sorts of other orbiting
bodies in and out of our own solar system, which is to say
the moons of the various planets in the solar system and
now exoplanets orbiting other stars in our galaxy—"Before
this, it was as if we were botanists with just one flower to
study. But now it's as if we have been given access to an
entire garden of different kinds of flowers"—such that the
pertinent question becomes, how well do all those orbital
systems conform to the theory? We already know that of the
ten planets orbiting the sun (including the minor planets

Ceres and Pluto), nine do conform within a couple percent of the Bode-predicted orbit. Indeed, setting aside Neptune, it turns out that all the various divergences from Bode of the other nine planets (percentages of 101.18, 96.36, 101.26, and so forth) average out to precisely 100 percent. "Let's just set Neptune aside for a moment," Murch suggests; "it's a renegade, so we will put it in detention until we can find out why it is behaving badly" (whereupon he launches a table of such outliers).

But what, for instance, about Jupiter and the other gas giant planets? Jupiter, as it happens, has sixty-three moons, though the vast majority are less than a few kilometers in diameter, so for the purpose of "weeding out the garden," Murch says he will limit himself for the moment to moons larger than 200 km in diameter, which in Jupiter's case leaves just the four moons discovered by Galileo: Io, Europa, Ganymede, and Callisto. (In passing, Murch notes that these four moons of Jupiter are roughly the same size as our own moon—which is to say our own Moon is much, much larger, relatively speaking, compared with its planet than Jupiter's moons are compared with theirs.) Of those four (Galilean) moons, the three outer ones (Europa, Ganymede, and Callisto) fit Murch-Bode to remarkably close degrees (percentages of 100.3, 99.71, and 100.27 respectively) and also, as we saw at the outset, overlap almost exactly, relatively speaking, with the orbits of Earth, Mars, and Ceres. The innermost moon, Io, however (inside of what would have been the corresponding Venus orbit), is the outlier (or should we say inlier?) here, and gets temporarily put in detention alongside Neptune.

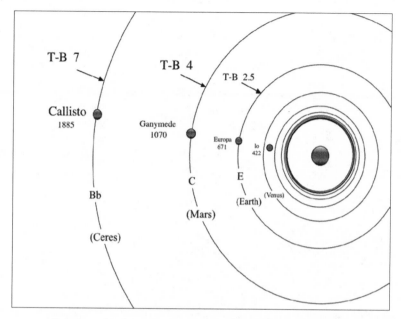

Relative orbits of Jupiter's moons and the Sun's inner planets

Saturn has nine moons larger than 200 km, and five of them follow Murch-Bode. Two of those, Dione and Rhea (at 100.96 percent and 98.70 percent of their predicted orbits respectively) slot into orbits like those of Venus and Earth; there is no match for Mars, but Hyperion (at 99.06 percent) rhymes with Ceres, and on the other side, Tethys and (closest in) Enceladus occupy orbits (to 100.33 percent and 101.88 percent accuracy respectively) previously unseen in the solar system (at n = −3 and n = −5, which is to say between Venus and Mercury, even though there is no equivalent of Mercury in the Saturn system). "This, I think, is particularly significant," says Murch. "The central differences from the original Bode theory are all these hypothesized orbits with negative values of *n*, inward of Venus's *n* at −2, as it were, and

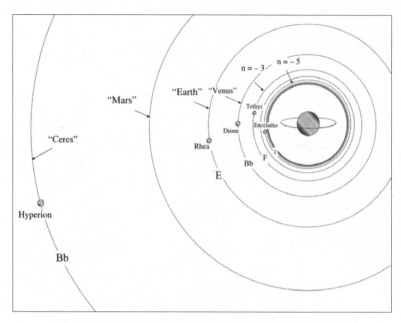

Relative orbits of Saturn's moons and the Sun's inner planets

here in Saturn's moon system we are seeing them populated for the first time—this having been precisely the wager we made when we extended the values of n into the negative infinity, and here we are seeing it pay off. Notwithstanding which, four of Saturn's moons (Mimas, Titan, Iapetus, and Phoebe) don't fit Bode and get sent to detention.

"As you can see," Murch is likely to quip around this point in his talks, "it's as if a virus had taken root in my brain, exhorting, '*You must complete me!*'" Three of Uranus's five moons larger than 200 km in diameter (Ariel, Umbriel, and Titania), Murch continues, slot into the Venus-, Earth-, and Mars-relative orbits (again to high degrees of predictive accuracy, with Titania at precisely 100 percent), and the fourth, Miranda, fits another of the hypothesized "unoccupied, inside Venus orbits" ($n = -4$), also at exactly 100 percent. Oberon is the renegade here and gets added to the detention list.

Neptune's moons are generally smaller than the moons of the previous three planets, so Murch lowers the criterion to 100 km in diameter. There are six of Neptune's moons to consider, three of which (Despina, Larissa, Proteus) occupy Venus, Earth, and Mars slots within a percent or two of their predicted orbits, and one, Nereid, which is way out in the distance, within 3 percent of the orbit predicted for $n = 6$, which is to say even beyond the relative orbit of Pluto ($n = 5$). Nereid's is the most elliptical orbit in the solar system, and the moon itself is quite irregularly shaped, both factors of which constitute good clues that it is a captured object: "With Neptune we are on the inner edge of the Kuiper Belt, another zone of millions of asteroid-like objects, and Nereid appears to have been one such that drifted in too close to Neptune, but the fact that even it seems to be conforming to Bode predictions suggests that Bode is a *dynamic* system, such that an object that gets sucked into any gravitational system later than that system's original formation nevertheless over time gets drawn into these Bodean undulations, so it's very provocative—both that Nereid's orbit is so far out and so elliptical and that its average distance nonetheless slots into one of the predicted orbits in Bode's law." (Two of Neptune's moons, Galatea and Triton, however, get sent to detention.)

Mars for its part has two small misshapen moons, Phobos and Deimos, each only about 20 km wide, and hence in all likelihood captured asteroids, and yet they, too, have fallen

Planet/Moon	n	% T-B Agreement	% Absolute Deviation
Sun			
Mercury	−∞	97.92	2.08
Venus	-2	104.56	4.56
Earth	(-1)	101.18	1.18
Mars	0	96.36	3.64
Ceres	1	99.79	0.21
Jupiter	2	101.25	1.25
Saturn	3	96.50	3.50
Uranus	4	99.07	0.93
Pluto	5	103.08	3.08
Mars			
Phobos	−∞	99.97	0.03
Deimos	(-1)	100.03	0.03
Jupiter			
Europa	(-1)	100.03	0.03
Ganymede	0	99.71	0.29
Callisto	1	100.27	0.27
Saturn			
Enceladus	-5	101.88	1.88
Tethys	-3	100.33	0.33
Dione	-2	100.96	0.96
Rhea	(-1)	98.70	1.30
Hyperion	1	99.06	0.94
Uranus			
Miranda	-4	100.78	0.78
Ariel	-2	100.89	0.89
Umbriel	(-1)	98.52	1.48
Titania	0	100.79	0.79
Neptune			
Despina	-2	101.43	1.43
Larissa	(-1)	99.35	0.65
Proteus	0	99.21	0.79
Nereid	6	96.40	3.60

Average Absolute Deviation   1.37 %

*Twenty-seven out of thirty-six Solar System objects
which fit Titius-Bode to a tolerance of a few percent or less
with an Average Absolute Deviation of 1.37%*

into Bodean troughs, Phobos matching Mercury's negative infinity orbit at 99.97 percent and Deimos matching Earth's at 100.3 percent.

Dropping the criterion even further, Saturn and Uranus have a whole raft of smaller inner moons, almost all of which line up with those negative−*n* (inward of Venus) orbits to a remarkable degree.

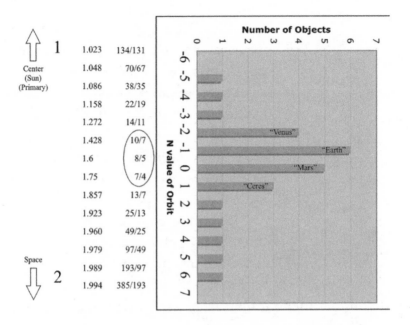

The ratios between adjacent orbits and the number of objects in each

"If we look at the big objects collectively," Murch summarizes, "twenty-seven out of the thirty-six fit Bode to a tolerance of four percent or less, many much more accurately.

"Furthermore, in *every* system there is an object occupying the so-called Earth-equivalent orbit, where n = –1. Indeed, there is a bell curve leading up to and away from *n* at negative one (Earth), such that there are all sorts of Venus-, Earth-, Mars-, and Ceres-type orbits, with progressively fewer to either side of those." (This may have something to do, he goes on to hypothesize, with the fact that the ratios between those orbits can be represented by simple fractions, $10/7$, $8/5$, and $7/4$, whereas the fractions representing ratios between adjacent orbits get considerably more

Neptune
Io
Mimas
Titan
Iapetus
Phoebe
Oberon
Triton
Galatea

The Nine Renegades

complicated the farther to the sides of the bell curve one goes: for example, the ratio between $n = 6$ and $n = 7$ is $385/193$.)

But what, Murch wonders, are the implications of the "Earth orbit," where n is -1, being super-fertile? "I didn't expect to come across this observation," he confesses, leaving the question hanging, a mystery.

Returning to the renegades, though, "the Bad Boys of the class" as Murch now characterizes them: it turns out, upon further investigation ("Is there anything that unites them in their badness?") that five of the nine (Neptune, Titan, Io, Oberon, and Galatea) find themselves almost exactly halfway between two predicted (though not necessarily occupied) orbits to either side.

"What is significant about these midpoints," Murch continues, "is that they are simply a reiteration of the near-most inner orbit, rather than a doubling of it. It is as if there were a kind of stuttering." As for the four remaining renegades

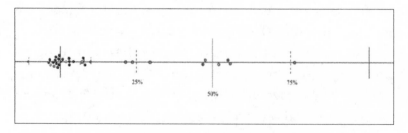

The relative distribution of the Renegades
in relation to predicted orbits to either side

(Mimas, Iapetus, Phoebe, and Triton), they turn out, while not fitting the midpoint, all to fit the midpoint of the midpoint of where they are supposed to be (the one-quarter and three-quarter marks), though Murch acknowledges that "Here, admittedly, we are drifting into thin oxygen, conceptually."

Meanwhile, over the past couple of decades, especially thanks to the more exacting exposures afforded by the orbiting space telescope named after Kepler, hundreds of exoplanets have been discovered orbiting nearby stars. Indeed, so many, and more so with every passing week, that compiling any sort of completely up-to-date statistical inventory is an almost hopeless undertaking. Still, when Murch last looked, a couple of years back, 24 out of 38 two-planet systems fit Bode (with a majority of those having planets in the Earth/Mars relative orbits); even HW Virginis, a rare double star, sported a pair of planets in a Venus/Earthlike configuration, and 8 out of 11 three-planet systems did so as well. Here again, a sizeable plurality of the planets in question seemed to fall into the fertile Venus/Earth/Mars/Ceres orbits, relative to their own systems. Indeed, planet b in the HR8799 system, which Murch cited at the beginning of his talk, coincided with the relative Earth orbit at precisely 100 percent.

Still, it was only at the beginning of 2016 that things really began to change for the scattered die-hard proponents of Titius-Bode. For as Murch now informed his audience, early in February 2015, Charles Lineweaver, a Berkeley Ph.D. now an associate professor in the Australian National University's College of Physical and Mathematical

Lineweaver   *Bovaird*

Sciences (with an almost uncanny facial resemblance to Johann Titius), and his graduate student Tim Bovaird published a peer-reviewed article in the *Monthly Notices of the Royal Astronomical Society* entitled "Using the Inclinations of Kepler [Telescope] Systems to Prioritize New Titius-Bode–Based Exoplanet Predictions." As Murch went on to note, "For outsiders to the field, they might shrug, 'Well, okay, so what,' but in astronomical circles, to get the words *Titius-Bode* into a serious paper in a peer-reviewed journal is a real breakthrough, especially a paper in which the authors claim that 86 percent of the exoplanets in systems with four or more orbiting bodies fit the Titius-Bode forecasts, and predict that further researchers should be able to find a planet here or here or here." (Murch noted that Lineweaver and Bovaird have tinkered with their Titius-Bode formulation in a novel manner of their own, with certain fascinating implications—to wit, that each system may have its own compounding multiple, such that though our solar system seems to abide by a doubling principle, other systems might compound at 1.7 or 2.3, but once you figure out the exact number, the system conforms to Titius-Bode, and likely predictions can therefore be advanced.)

At this point, Murch alluded to the way that Dmitri Mendeleyev deployed similar predictions when first exhibiting his Periodic Table of Elements, likewise alerting colleagues that they should focus their search for new elements here and here and here, at these holes in the table. Lineweaver and Bovaird had issued their prediction several months before this, and had already been claiming a 5 percent hit rate for planets in fact showing up (in the empty spaces they had identified in the 86 percent of exoplanet systems that they had shown to be following their version of Titius-Bode), which, as Murch elaborated, "depending on your point of view is either very low or very high." Indeed, a pair of astronomers out of Cornell had already attacked them, saying that 5 percent is not much better than chance, and this piece in the *Monthly Notices* was in part Lineweaver and Bovaird's response to that accusation.

Seeing exoplanets, especially small ones, with the Kepler telescope turns out to be extremely difficult: one sees them only if they happen to cross in front of the star in question, thereby dimming the light that the Kepler telescope is receiving from that star during the relative pinprick's transit, from a careful study of which all sorts of characteristics can be inferred—mass, distance from the star, rate of the orbit, and so forth. But that works only if the planet's often-quite-long orbit happens currently to be on this side of the star and then, furthermore, happens to traverse the extremely narrow plane between the star and the face it turns to the Kepler telescope, not below and not above, and beyond that is big enough to be seen—planets the size of Mars and Mercury, for instance, wouldn't be— and what are the chances of that? Thought about that way, 5 percent might seem like a pretty good rate, and in fact, in the meantime, Lineweaver and Bovaird were claiming in this new piece an increase to 15 percent. "They are getting all sorts of pushback," Murch acknowledged, "we are still dealing with the default cooties factor here—so this is a battle of the bands now, but what's fascinating is that for the first time in a hundred seventy years, the battle is being waged on Titius-Bode turf, which is a big step forward, as far as I am concerned."

FINE, I SAID to Murch over dinner later the evening of the talk, but what about Gauss's third objection: what did all of this mean, what could possibly account for it? *L'univers*, in short, *a-t-il un sens?*

"Ah, yes," Murch acknowledged, chuckling, "time ran out on me during the lecture before I could get to that. And frankly, here I am on the thinnest ice—I mean, I am not an astrophysicist, and I can only speculate in the most tentative way. And again, I am primarily interested in presenting the data and getting people to take another considered look at Titius-Bode armed with all the data that we now have. Nevertheless, naturally, as in any dance between evidence and hypothesis, I've given the matter some thought.

"Because the fact is that what we are seeing are unexpectedly organized systems, more so, it seems, than can be explained by conventional Newtonian or even Einsteinian models. And for me it comes back to the analogy to music:

once I saw how Bode ratios dovetailed so perfectly with the octaves and sequences of notes in just intonation, I naturally made the association to sound waves, which in turn got me to thinking in terms of undulations, of waves generally—so maybe we are seeing something like that."

Actual waves? Perhaps even gravitational waves at that?

"Well, perhaps. I don't know..."

Murch's comment reminded me of a passage I'd read the night before, boning up for his lecture, in Lincoln Barnett's classic *The Universe and Dr. Einstein,* and pulling the book out of my satchel, I now tried it out on Murch:

> The distinction between Newton's and Einstein's ideas about gravitation has sometimes been illustrated by picturing a little boy playing marbles in a city lot. The ground is very uneven, ridged with bumps and hollows. An observer in an office ten stories above the street would not be able to see these irregularities in the ground. Noticing that the marbles appear to avoid some sections of the ground and move toward other sections, he might assume that a "force" is operating which repels the marbles from certain spots and attracts them toward others. But another observer on the ground would instantly perceive that the path of the marbles is simply governed by the curvature of the field. In this little fable Newton is the upstairs observer who imagines that a "force" is at work, and Einstein is the observer on the ground, who has no reason to make such an assumption. Einstein's gravitational laws, therefore, merely describe the field properties of the space-time continuum.

Was that the sort of thing Murch had in mind?

"Actually, yes, but remember that we are speaking 'as-if,' in Vaihingerian terms," Murch said, invoking a favorite philosopher of both of ours, the great neo-Kantian Hans Vaihinger, author of *The Philosophy of "As-If"* (1911, though written thirty years earlier). "It is that things are behaving 'as-if.' And what I am suggesting is that one might reimagine

Vaihinger

the standard Einsteinian model of the warpage of space-time in the presence of massive objects—you know, the metaphorical image of a bowling ball on a trampoline—to suggest that rather than there being a smooth hyperbolic sag, instead things are behaving *as if* there is also some kind of rippling wave texture to it, following the Bodean-Pythagorean sound wave model, starting at beta and moving outward from the edge of the smooth, unrippled cordon sanitaire around the gravitational center, and that orbiting planets and moons tend (at least eighty percent of the time) to fall into the troughs, so to speak, between the peaks of those ripples. And while we haven't yet found a way of seeing or measuring these hypothetical ripples themselves, we are able to infer their existence, statistically, through the way they seem to work on planetary bodies revolving around a common center."

As with the Robert Bresson line from his fortune cookie: "Translate the invisible wind by the water it sculpts in passing."

"Well, you can see why I like that one. But let's clarify what is meant by 'waves' here. There appear to be zones or valleys of lesser pressure where things collect and others, peaks of higher pressure, where they don't, as in a standing wave—exactly the same as with sound waves in an enclosed space, which is always the challenge in designing concert halls or movie theaters."

What did he mean by "standing" waves?

"In the ocean, so-called *propagating* waves move all the way from Japan, say, to California, eventually crashing onto the shore at Bolinas. These in turn might be seen as analogous to the sort of gravitational waves, more conventionally understood, that astrophysicists theorize might be welling out, at the speed of light, from collapsing black holes and the like, millions or billions of light-years away, the kind of thing whose undulations, almost infinitesimally small by the time they reach us, such technicians are looking for by way of billion-dollar high-precision experiments. When it comes to ocean waves, though, the actual particles of water along the way more or less stay in place, bobbing up and down, while the wave moves *through* them. Whereas, in a *standing* wave, it's the opposite: the water moves, and the wave stays in place. Imagine a boulder in the middle of a fast-coursing river: the water on the upstream side of the boulder piles up behind the boulder, and a series of standing waves forms on that upstream side: only, this time it's the particles of rushing water that move into and through the waves, which in turn stay in place, and are therefore said to be standing.

"This is all a bit reminiscent of those famous Chladni figures," the phenomenon named after the eighteenth-century

German physicist and musician Ernst Chladni, who first elaborated them (thereby earning the epithet "the father of acoustics"): the sort of thing that happens when you draw a bow along the rim of a flat metal plate that has been sprinkled over with sand or flour or couscous until the plate achieves resonance, at which point the vibrations cause the sprinkled particulates to shift and concentrate along nodal lines where the vibrations cancel out and the plate is not moving. The effect can indeed seem quite magical.

Chladni and his figures

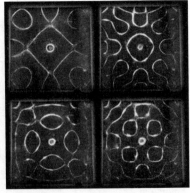

"For that matter, there's an amusing thing you can do with ants in a microwave oven," Murch continued. "If you place a bunch of live ants scattered about a stationary platter into a microwave, close the door, and turn the oven on, you will see—don't worry, no ants get harmed in this experiment—that the ants will self-organize into patterns, quickly moving away from where the microwave energy is highest and toward where it is less so. That's why they had to put rotating plates in microwave ovens: before that, meat cooked quite unevenly, raw in some places, burnt in others. I'm suggesting something similar may be going on with orbiting bodies which migrate, as it were, to zones of less 'pressure,' which in turn seem to be welling out from the central body in a Bodean array."

Which reminded me, I'd meant to ask him: back during the talk, he noted that several of the renegades, the bodies in detention (Neptune, for example) turned out to end up almost exactly halfway between two troughs in the wave pattern. But wouldn't that have put them on the wave peaks, the zones of highest pressure and hence exactly where he was saying they could never go, like the ants?

"Well, the midpoint isn't exactly in line with the peak of the adjacent wave. If you work out the math"—he reached for a piece of paper and demonstrated what he was talking about— "It's easier just to sketch this out but you'll see that the midpoint turns out to be one eighth of the valley's width 'down the slope' from the intervening peak toward the second trough. Just to make things more complicated—or, as I see it, more interesting—if the successive peaks represent octaves of the note C, then the renegade is lodged at a hollow formed at D."

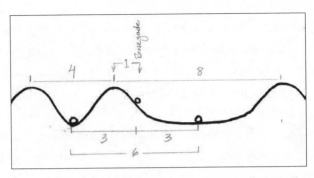

The paradoxical midpoint between two troughs

Okay, I now hazarded, but if you are getting a standing wave, as it were, that implies that something is flowing?

"Well, you would think so, wouldn't you? Something is flowing, as with a river, and backing up as if upstream from the boulder: the thing that is flowing is moving through the system up and down and up and down, as it were, while the wave itself stands in place."

Heraclites (master of flow), I suggested, meeting Democritus (conceiver of atoms).

"Indeed. But what could that be? Here of course we are way, way, *way* out on a limb, speculatively speaking. But what is flowing in Einsteinian time-space? Perhaps, precisely, time? Or else, maybe, all that dark matter? Alternatively, this is where we might need some kind of conceptual jujitsu move yet to be achieved. But the point is that the data is what it is, it's not a mirage, it is saying what it is saying, and it needs, at the very least, it seems to me, to be entertained, if not downright confronted."

PART TWO

Troughs and Swells

FEW OF THE astrophysicists I now began trying to speak with, however, were in any sort of entertaining mood.

Five years ago, while serving a stint as a visiting professor at Occidental College, in Los Angeles, I organized an all-day public Wonder Cabinet (a heterodox series of presentations of all sorts, just one cool thing after another), as part of which I had invited Murch to give the then-current version of his Titius-Bode talk, already a powerfully engaging PowerPoint presentation, whatever one might have thought of its ultimate veracity. As I was helping the speaker right after Murch, an eminent Caltech physicist with a side interest in the formation of snowflakes, to effect the digital transition to his own laptop, I inquired, sotto voce, what he had made of the preceding talk, and in particular what had seemed to me its pretty impressive marshaling of mathematical evidence. Without missing a beat, the physicist replied, "Numerology."

More recently, I asked another scientist friend of mine, Charles Falco, the distinguished quantum optical physicist at the University of Arizona in Tucson, whom I had asked several years ago to help advise on David Hockney's then-radical notions about whether Old Masters might have been using optical devices much earlier than had previously been thought (and who had proved exceptionally open to that possibility, indeed, eventually joining Hockney on several trailblazing papers) to take a look at a video of Murch's most recent iteration of his theory. From earlier experience, I was expecting Falco might be a bit more open than some of the more doctrinaire astrophysicists I'd been consulting. Far from it: "Okay," he replied in exasperation, by way of e-mail, "I took the time to listen. But I bailed after fifty-three minutes; it was just too aggravating to continue, and I already had two pages of notes of objections," which he thereupon went on to list. Among them:

- Ceres? Seriously? There isn't a planet where he needs one so he arbitrarily calls an asteroid a "planet" and incorporates it. But, later (where I gave up), he throws out all the moons of Jupiter that don't fit by arbitrarily defining them as "small." This is a fundamental problem throughout. He selects data he needs to make things sort of fit, and tosses out data that doesn't fit.

- He invokes Charles Lineweaver as supporting his concept. I Googled him, and it appears he is 15 years past his PhD and is still only an assoc. professor. Not to say he might not be a serious scientist, but listing him as "PhD Berkeley," rather than "Assoc. Prof." seems to be recognizing a weakness in the citation[…]

- He makes a big deal out of getting rid of the earth-centric AU as if that somehow is a limitation, but he uses the Western-centric octave scale.

- He draws visually compelling crests and troughs and places planets in the troughs, visually "proving" they are stable there. But, this is *totally* arbitrary and is visual smoke and mirrors.

And so forth, concluding, "Yours in Allah, peace and numerology be upon him." When I wrote back, asking Falco to reconsider some of his points (it seemed to me he had been a bit peremptory in his viewing of the video, perhaps not listening carefully enough to what Murch was trying to lay out, for example, in the case of the discovery of Ceres and the asteroid belt, which, after all, hadn't been Walter's contention alone but rather a central event in the growing acceptance of the original theory early on), Falco was slightly more circumspect in his reply ("There is no doubt whatever he's a fascinating person, and I would very much enjoy having a long conversation with him"), but he again concluded,

However, he has two strikes against him in this Bode Law business. First, he plays fast and loose with counting data when it's useful to him and discounting it when it's not. Second, it would mean that even though every test of gravitational forces over distances from sub-atomic to intergalactic supports a $1/r^2$ dependence (i.e. not waves), somehow none of those tests caught this astro-acoustical wave stuff.

Just because no one found something before doesn't mean it's wrong. But when something apparently new is found in a well-tilled, highly quantitative scientific field, the bar for skepticism doesn't allow for being selective with the data and invoking ad hoc explanations to fudge outliers into compliance.

When I e-mailed another scientist to whom I had been referred, in this instance an eminent astrophysicist named Richard Greenberg (who also happened to be at the University of Arizona, one of the foremost centers of optical and astronomical research in the country), with a summary of Murch's arguments (including Murch's own proviso about the hazards of apophenia), Greenberg initially wrote back:

> Well, it all sounds rather apophenic to me. I get lots of email from out of the blue with various people's discoveries of meaningful patterns in planetary data that explain life, the universe, and everything.
>
> Of course some patterns can be meaningful, but really they are only of value if they can be the starting point for finding the actual physical processes that were involved. So, for example, as data are accumulating from discoveries of extra-solar planetary systems, there is a lot of interest in looking for common patterns that might point to how planetary systems form. But identifying patterns is only a first step, and they won't mean much unless we can figure out what processes caused them.
>
> Another big issue involves the statistical significance of numerology. What may seem like a significant pattern, often is not. Similarly, while a pattern may seem to be predictive, it really needs to be statistically significant. A "98–102% degree of accuracy" does not sound like something

that someone understanding real statistics would say. I think that the one published paper that tested Bovaird et al.'s predictions found that they were not particularly convincing.

"The bit about gravitational waves in the space-time continuum affecting the locations of planets sounds really nutty," Greenberg concluded. "It makes analysis of the Titius-Bode law seem sane by comparison." Still, Greenberg agreed to give the video a look, though, in the end, he was hardly swayed from his initial suspicions. In a second e-mail, he began by noting how

> astronomers do try to look for patterns in data, and there is plenty of that going on now, especially with the many newly discovered planetary systems. But where they differ from Murch is that the whole thrust is toward finding the physical processes that create these patterns. Murch looks amateurish because, well first he is an amateur, but more fundamentally because the numerology seems to be an end in itself. There is no effort to test for statistical significance, nor to consider the responsible physical processes. The only thing close to a physical explanation is the purple wave pattern [*see page 52*], but that is hardly explained and just seems kooky.

"Murch comes off as a charming amateur, who is having a good time playing with numbers, but there is nothing new or profound in what he is finding," Greenberg added, before graciously concluding, "I imagine I would look even more amateurish if I tried to edit a film! Murch is probably much better at his profession than I am at mine."

Lawrence Krauss, another celebrated astrophysical theorist to whom I had been referred (also in Arizona, though at Arizona State University in Tempe, where he is director of the Origins Project and codirector of the Cosmology Initiative) was yet more definitive in his evaluation. Without bothering to watch the video, he responded to my summary of Murch's ideas with an emphatic "Claiming some relationship to gravitational waves is simply nonsense. The gravitational waves emitted by planets orbiting stars, or normal stars themselves, have an amplitude which is simply undetectable, and can have no physical effect on the orbits."

Over the months, as these sorts of responses came in, I took to forwarding them to Murch, who was now in London, working on an Iranian émigré's new documentary on the 1953 CIA coup that upended the democratically elected government of the Persian nationalist Mohammad Mossadegh. (Talk about wave upon wave of consequence, and consequences of consequences!) Murch was invariably patient, thorough, and exact in his rejoinders.

Thus, for example, on the specific criticism that he had cherry-picked his numbers, he protested that, on the contrary, he had been quite scrupulous in detailing the "renegades" that didn't fit the general model and in trying to account for what they were perhaps doing instead. What about Falco's objection that he, Murch, had limited his analysis to the four biggest moons of Jupiter, ignoring the other fifty-nine?

I confined myself to moons greater than 200 kilometers in part because that's the diameter above which astronomical bodies generally achieve spherical shape; below 200 kilometers, they are most often irregular, since their small mass isn't enough to force them into a sphere. But more to the point, I was just trying to get an initial approximate handle on the underlying landscape, like setting tolerances on a topographical relief map. And the preponderance of those other fifty-nine are really, really tiny: if *all* of them were compressed into a single object, that object's mass would amount to only $1/2000$th that of Europa, the smallest of the moons discovered by Galileo. Now, without question there will be many individual asteroids or minuscule moons that don't fit Bode, but those are kind of like dust motes in the air, where the random factor is high. As it happens, I did run the numbers on the first 200 asteroids in the order of their discovery—recall that Ceres, the first to be discovered, accounts for $1/4$ (at least) of the total mass of the asteroid belt, and that the next 199, by definition probably the next largest because the most visible, amounted to over 95 percent of the total mass, even though there are (at last count) over 2 million asteroids in the belt, most of them Little Prince size or less—but the point is that the common center of those first 200 asteroids fit Bode with almost 100 percent accuracy.

What of the more general criticism that he was just playing with numbers? "What can I say," he wrote back,

They're right, I am having a good time with numbers, but number-play is an important part of human intellectual effort. The question is: does the number-play link to physical realities? And it seems to me that there is something going on with Bode that at very least warrants further investigation. I haven't made a deep effort to test for statistical significance because that is outside my skill set. I had hoped over the years that I could meet someone who might be able to help me with this, and I've frankly been exasperated at my inability to rouse such interest.

On the other hand, he went on to note that he was frequently coming upon scientifically validated (and, more to the point, accepted) principles that rhymed uncannily with the sort of thing he was noticing. Thus, for example, Murch sent me a recent piece from *Symmetry Magazine* (September 2014) entitled "What Hawking Really Meant," on the current consensus around the concept of metastability:

Now we need to bring in the universe and the laws that govern it. Here is an important guiding principle: The universe is lazy—a giant, cosmic couch potato. If at all possible, the universe will figure out a way to move to the lowest energy state it can. A simple analogy is a ball placed on the side of a mountain. It will roll down the mountainside and come to rest at the bottom of the valley. This ball will then be in a stable configuration.

The universe is the same way. After the cosmos was created, the fields that make up the universe should have arranged themselves into the lowest possible energy state.

There is a proviso. It is possible that there could be little "valleys" in the energy slope. As the universe cooled, it might have been caught in one of those little valleys. Ideally, the universe would like to fall into the deeper valley below, but it could be trapped.

This is an example of a metastable state. As long as the little valley is deep enough, it's hard to get out of. Indeed, using classical physics, it is impossible to get out of it.

If such a thing occurred at that scale, Murch wondered, why couldn't one imagine something similar happening at the level of lunar or planetary orbits, along the interior slopes of the Bodean waves? "What is without question true, though, mathematically, is that the Bode formula generates musical ratios, and insofar as the Bode formula describes the relative positions of the solar system's planets (and their moons) it also describes a musical configuration."

But what, then, about the criticism that he was imposing a Western-centric musical model on the cosmos?

No, he replied. If he had been citing the so-called "well-tempered" tuning of the sort that Bach, for example, innovated and explored, that would be one thing.

But the musical scale I am describing is rather what we might call "the music of nature," and it is far from a human contrivance, Western or otherwise. Every object when struck, bowed, plucked, or blown vibrates at its "dominant" frequency based on its shape and size and composition. The best example might be a bell. But a string, or a hollow tube, a wheel from an abandoned automobile, etc.

all do the same thing. And, they also vibrate at integral multiples of this dominant frequency (2x, 3x, 4x, 5x, 6x etc.). Each subsequently higher vibration (harmonic overtone) has less energy than the previous, following approximately the law of inverse squares ($1/4$, $1/9$, $1/16$, $1/25$ etc.). The "approximately" varies with the exact tensile strength and composition of the object in question. And it is the exact blend of overtones that tells us whether the object

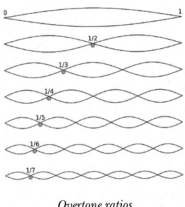

Overtone ratios

in question is a violin, flute, bell, or old wheel. Also, the trained ear can distinguish a Stradivarius from an inexpensive 19th-century violin using the same comparison of overtones. Now, if the dominant note is C, then the overtones are 2C, 3C, 4C, 5C etc. and we call these overtones, in our scale system: C, G, C, E, G, Bb, C, D, E etc....But they would "be there" in nature (*are* there in nature) whether we call them anything or not.

And was he the first to have noticed the overlap of those scales with Bode? He believed so—it was, at any rate, noticing that, he said, that had gotten him going on his whole notion of waves with barren peaks and fertile troughs.

Why did he think no one before him had ever noticed this identity of Bode and music?

"I don't know...Kepler certainly might have, but the Bode relation wasn't known then, though he himself could easily have discovered it. And yes, by the time Bode came along (150 years after Kepler), the idea of the music of the spheres had pretty much faded from the culture."

Rob Speare, an NYU cosmology grad student I'd struck up a conversation with at one point, critiqued Murch's analogy by pointing out that musical waves, as in Murch's own example, require bounded conditions (the string, for instance, had to be bounded on both sides), which was not the case with a solar and planetary system, which was only bounded on the one, at the gravitational center. Murch, for his part, responded that this was true for the acoustics of musical instruments, but not for the example of the standing waves upstream from a boulder in a river.

A few days later, I pointed out to Murch how, speaking of those astro-acoustic waves, virtually every physicist I was polling was adamant that whatever it was he was noticing, they couldn't be gravitational waves. As that NYU cosmology grad Speare, for example, had just written me:

> The Titius-Bode coincidence/law *definitely cannot* be explained by Gravitational waves. To a physicist, Gravitational waves mean perturbations in the *metric*, the way we measure distance between points in space time, and that needs to derive from a *very, very* strong affect. Gravitational waves are only generated by extremely large objects—like Black Holes—slamming into each other at relativistic velocities or the Big Bang. (Even from those sources, signals are tiny.) A forming Galaxy would not be able to generate enough energy to make those kinds of waves.

Again Murch responded by citing a rhyming passage in a recent article, this time in the *Economist* of August 22, 2015, on the current state of play on thinking about dark matter—to wit, that "Light and matter pushed one another to and fro in the early universe, creating pressure waves. As the early universe expanded, matter was scattered along these so-called baryon acoustic oscillations, more densely packed at the peaks of the waves and less so at the troughs." Now, Murch continued, acknowledging Speare's point about a distinction in kind and scale,

> obviously they are talking about fluctuations on the galactic or extra-galactic level, but at least it shows that they are speaking with the same terms as I (we) have been: baryonic acoustic oscillations, with greater density at peaks and less so at the troughs of these oscillations.
>
> I'm sure (?) the people you've been in touch with would say that the force of a planet in orbit is greater, much greater than the force exerted by dark matter. And that would be true, I would add, if you were talking about hundreds, or thousands or even hundreds of thousands of years. The force in question is indeed weak, very weak, and wouldn't make itself felt in normal time perspectives (the way we don't see tree growth with our normal mammalian time-sense). But my hunch says that we are talking about millions and hundreds of millions, billions of years, where the relentless constant influence of the dark matter or whatever it is is a steady drip drip drip, nudge nudge nudge.

At which point, Murch launched into an analogy of his own:

> Thirty years ago, I was on a solo expedition, looking for locations in Kansas for the *Return to Oz* film I was about to start directing, and I would break into long-abandoned farmhouses, buildings that had gone undisturbed for twenty and more years, and I'd notice how over that length of time, tiny tiny grains of very fine dust had settled into the undulations in the wood grain of the floorboards. Think of how small the force of gravity would have been that was exerted on those tiny grains, and the tiny force exerted by the slope of the undulations. But with enough time, they worked their magic.

He concluded with a wry parenthetical "(Am I allowed to use the '*m-g-c*' word? Probably not...)"

One time over dinner, Murch pushed his chair back in exasperation. "The question as to *why* there are these Bodean intervals is the readily available big stick that is regularly used to break the noggins of people like me who think about these things; and indeed, in terms of concrete current understanding, they have a point, but we are precisely trying to move beyond current understanding."

Another time, more recently, I wrote Murch in London to ask if he had ever heard of Edgar Allan Poe's screed *Eureka*—there happened to be an exhibition going by that title at the Pace Gallery in Chelsea whose curation had been inspired by the monograph, and on one of the show's wall panels, I'd found Poe citing Bode—and of course, naturally, he had. "Yes," Murch replied instantly, almost exultantly:

Poe

Eureka! I read it (in an Italian translation!) in 1995 when I was in Rome on *The English Patient*. Poe did write in 1848 about Bode's law, when the news of Neptune's discovery (1846) and its violation of Bode's law had already begun to spread but the implications had not yet been fully absorbed. Certainly not by Poe.

It is a fascinating work—Poe believed it was the most important of anything that he had written, and he dedicated the last years of his life to its promulgation. You might even say that the effort of promulgation contributed to his death, so all-consuming was it.

Of course many things are wrong with *Eureka,* from our perspective, but if you read it with the right pair of glasses, you can see—struggling to be born—the idea of the Big Bang: the instant of the creation of the Universe ex nihilo. As he memorably put it: "Because Nothing was, therefore All Things are." If Einstein, sixty years later, rebelled against his *own theory's* implications of an evolving universe (inserting the cosmological constant to stabilize it), you can imagine how *Eureka!* was received in the middle of the 19th Century. Particularly coming from someone like Poe, a threadbare writer of mysteries and horror stories.

Murch seemed to identify with Poe.

Other times, Murch would say, "Of course I am speaking analogically or, more accurately, in a Vaihingerian manner," again invoking the neo-Kantian philosopher behind *The Philosophy of "As-If."* "The model I am seeing in these orbital systems behaves *as if* there were weak standing wave undulations in the density of space-time (or at any rate standing wave undulations in *something*)."

"I mean," he would sometimes elaborate, "*I know I'm wrong.*" That wasn't news to him, that some of his ideas, especially with regard to causalities, weren't quite fitting together and could easily be shot down as such. Something was missing, or perhaps needed flipping in some as-yet-inconceivable (to him or anyone else) way. But that, as Vaihinger had taught, is how science works: stumbling forward by a sequence of analogical, as-if approximations—useful "heuristic fictions."

One night, over a meal of sushi in Manhattan, on one of Murch's passes through the city between London and Bolinas, he offered up (by way of analogy of course) a history lesson, starting with his hero: "Kepler had only a dim idea of what made the solar system tick (he thought it was magnetism, not gravity), but he did get the idea (and he was the first to do so) that there were two opposing forces at work: the Sun was exerting some kind of force, and the planets were somehow resisting it. Everyone before him had thought there was only one force: angel power—the planets were being pushed around their tracks by divine impetus. That is what made the idea of ellipses so disturbing to everyone (including Kepler): why would angels push the planets in ellipses rather than (perfect) circles? Everyone, including him, said that was ridiculous, and he characterized it as a 'cartload of dung' right there in the middle of his theory. But when he at length came to terms with the elliptical nature of orbits, he put it down to the balance of forces occasioned by the way the sun sent out magnetic spokes which were fast in close but slower further out. He was wrong in that it was not magnetism, but he was right in that he saw a force welling out from the Sun that obeyed inverse square laws and a resistance at the other end. Newton would reframe the phenomenon in terms of inertia and momentum, and cross out the 'magnetism,' replacing it with 'universal gravitational force.'

Kepler   *Newton*   *Einstein*

"Kepler hadn't known what it was, but he just knew that if you accepted the notion—in what we can describe today in Vaihingerian 'as-if' terms—you could generate exponentially better data, such as the Rudolphine tables that Kepler produced in 1627, and which continued to be used well into the mid-nineteenth century. Newton didn't need to do any fresh calculations; he just piggybacked on Kepler in arriving at his jujitsu breakthrough, looking at Kepler's numbers in a whole new way and thereby calming things down, but he punted as to *why* it worked, not caring about why it worked—*hypotheses non fingo*—only about *how* it worked. *How* could be useful, though: you could use it, for starters, in calculations involving better ways of killing people with artillery shells.

"Einstein finally explained why it was true—gravity being not so much a force as a distortion in the space-time field—but then quantum effects arose, and they were back to 'why' all over again. At Copenhagen, Bohr, among others, said it's useless to try to figure out why, 'Just shut up and calculate.' And indeed, because of that we have GPS and drones, and yes, all sorts of better and better ways to kill people."

Murch paused for a moment, stabbing one last dab of sushi before continuing: "Maybe the problem with Bode is that no one has yet figured out a way of instrumentalizing it for the purposes of killing people. Though it continues to be useful as a teaching tool in astronomy classes in the mode of aversion therapy: '*Don't think this way!*'"

Murch motioned for the waitress to refill his tea cup and now palmed it, waiting for the tea to cool. "It's a bit like the story with the phlogiston effect. Joseph Priestley was an amateur scientist back in the eighteenth century who studied what happened with plants when you put them under a bell jar. He noticed that when you put a mouse in with the plant, the plant lived, and when you didn't, it died. He just got the reason for the effect backward, because

Priestley

he didn't know about oxygen. He imagined that when matter burns, a firelike element within the combustible object which he called phlogiston got released, and the burned-up thing, as it were, got dephlogistinated. Antoine Lavoisier came along and said, no, that when something burned, counterintuitively, something was being *brought in*: oxy-

Lavoisier

gen. The point is that Priestley had been right in every way except he had everything upside down."

Murch took in a final satisfying draft of tea. "And I may be doing something like that," he concluded. "Maybe it will turn out Murch was wrong, it's not dark matter or ripples in space-time, it's X. But the structure, I'm sure, is right."

A FEW DAYS later, Murch e-mailed me a passage from Maurice Maeterlinck—naturally, who else might one expect him to be reading?—from the latter's *The Life of the Bee* (1901, a decade before Vaihinger):

> Nevertheless, when it is impossible to know what the truth of a thing may be, it is well to accept the hypothesis that appeals the most urgently to the reason of men at the period when we happen to have come into the world. The chances are that it will be false; but so long as we believe it to be true, it will serve a useful purpose by restoring our courage and stimulating research in a new direction. It might at the first glance seem wiser, perhaps, instead

of advancing these ingenious suppositions, simply to say the profound truth, which is that we do not know. But this truth could only be helpful were it written that we never shall know. In the meanwhile, it would

Maeterlinck with bees

induce a state of stagnation within us more pernicious than the most vexatious illusions. We are so constituted that nothing takes us further or leads us higher than the leaps made by our errors. In point of fact we owe the little we have learned to hypotheses that were always hazardous and often absurd, and, as a general rule, less discreet than they are today.

I e-mailed Murch right back, asking how he had ever come upon *that*, to which he replied, "I was in Buenos Aires, and it was in a tattered Pocket Book reprint of *Life of the Bee* from the 1950's, fragile and yellowed, which originally sold, as they all did back then, for 35¢. A love of bees, I guess, drew me to it. From Walrus Books in San Telmo."

I then went and looked Maeterlinck up on Wikipedia (Belgian playwright, poet, and essayist, a Fleming, who wrote in French, who won the 1911 Nobel Prize for Literature) and came upon this quote: "At every crossroads on the path that leads to the future, tradition has placed 10,000 men to guard the past." Candidate for another Murch fortune cookie if ever there was one. And I sent it to him.

Duffy

*

I FINALLY FOUND an astrophysicist willing to engage Murch (and, by this point, me) at some length: Alan Duffy, the friend of a friend, and a noted professor at the Swinburne University of Technology in Melbourne, Australia. I sent him my usual packet of background material and a Vimeo link to the video of that most recent iteration of Murch's talk, and a few weeks later (apologizing for the delay by saying that he had had to wait till the end of the semester to be able to give the material the attention it deserved), Duffy replied by e-mail:

> I've had a chance to read Walter's brief notes and watch the entire Vimeo talk so I'll answer your questions in turn but first I'd like to say few words more generally.
>
> Firstly I was struck by how deeply intelligent and thoughtful a man Walter is, this much is clear from his presentation. In addition his amount of work on this topic is impressive and I would congratulate him on this as it's no mean feat to read so widely in this sometimes fairly opaque subject (thanks to nomenclature which, for a field as old as astronomy, can sometimes be archaic when it should be transparent and relatable, although Walter's classical education has undoubtedly helped here).

Walter's suggestion that the orbits from Titius-Bode (henceforth T-B) can be mapped to musical notes is interesting, and while I hadn't heard the musical link with T-B before, it's worth noting that any geometric progression (that is, a steadily doubling/tripling etc series) of which T-B is a single example, can be described as overtones.

Undoubtedly, Walter's training made that immediately apparent to him, but the link is in the very description of the underlying mathematics:

> Orbital motion is a regular pattern in time undergoing Simple **Harmonic** Motion...
> With Newtonian gravity you can expect certain Orbital **Resonances** (aka Mean Motion Resonances) to occur in specific ratios...
> for example we have 3:2 ratio of Neptune and Pluto (where Neptune orbits three times in the same time that Pluto has orbited twice).

Since Kepler's laws we have known that the period of the orbit is intimately tied to the size of the orbit, meaning there is a connection between resonances in period and the orbits of the same planets, so orbital resonances mean ratios can give these orbital ratios getting to the heart of the T-B result.

Why might this occur?

Imagine you're Pluto, orbiting outside a giant planet like Neptune...if you are lagging behind the planet, then Neptune's gravity will pull you along, boosting the speed and shortening your overall period; if instead you're leading the planet, then Neptune's gravity will pull back on you, slowing your speed and lengthening the overall period. This gentle pulling or tugging, repeated every time Neptune travels past you will cause your period to lock into some kind of ratio with that of Neptune's.

These mean motion resonances (MMRs) have been seen in the 1:2:4 ratio of Jupiter's moons of Ganymede, Europa and Io which Walter described (and I believe was noted by Laplace?) A ratio in period is a ratio in orbits... this is then the link to T-B.

Note that space is complex and many interactions will pull and tug a planet or a moon so that we don't expect the entire solar system to be in resonance (more likely things will be in resonance with that gravitational "bully" the gas giant Jupiter) *but the fundamental law is gravity, not T-B.*

It's worth noting that it's *extremely* complex to show if something like T-B drops out of how planets and moons form from gas disks around forming stars and planets, but the expectation is that simple Newtonian gravity (in an anything but simple situation like this!) will give that interestingly regular pattern without the need for extra physics.

I'd asked Duffy whether he credited Murch's raw numbers and the mathematical ratios that he generated from them, and he replied that he hadn't checked all the math but that the mean distances Murch quoted all seemed correct. I had gone on to ask whether he, as a practicing astrophysicist, had any problem with Murch's tolerance for orbits that came within 98 to 102 percent of the predictions, whether that constituted any sort of insurmountable statistical problem, and he replied,

As a talented musician Walter would be able to hear if an instrument was a few percent "off" its note, and this discrepancy is similarly jarring in gravity.

This is about the difference over a decade between Mercury's precession as predicted by Newton's laws and what is observed. One of Einstein's great initial successes with General Relativity was explaining just this sized effect in Mercury.

To put it another way, the only prediction of T-B (the orbital ratios) is not accurate enough for those orbits that it gets close to being called a Law, and in numerous cases (particularly Neptune) it is entirely wrong, which shows that it wasn't an underlying Law but rather an interesting pattern (perhaps a rule?)

We rejected Newton for Einstein due to just such a failure but, and here's the crucial point: we still use Newton's Law to predict the motion of tennis balls as for those purposes it's accurate enough. This is what led ANU's Prof Charlie Lineweaver to try and use T-B for exoplanet predictions. That this prediction was shown to be still too inaccurate by Princeton's Dr Chelsea Huang in 2014 is again another statement that T-B is not fundamental and worse yet, not accurate enough to be of much use either.

The fact that the sat-nav on your phone works is because the GPS systems account for Einstein's GR. The fact that we can get so close to a comet (within 100km!) after travelling 6bn km in one shot (akin to hitting a dartboard in New York with a dart thrown from LA) is because celestial mechanics and dynamics work so well.

Astronomers would love nothing better than to prove Einstein wrong—you'd probably get a Nobel Prize! The reason T-B has fallen out of favour is because it's not been able to predict so many systems. It can get pretty close sometimes, but that's no use if you're trying to fly a probe to Neptune as you'd miss it by a BILLION kilometres or more.

The link to music is a beautiful analogy but you get the same pattern of harmonic overtones (i.e. doubling) as T-B by counting the number of bacteria in a petri dish as they grow in time—doubling every 40 minutes or so for the strain *Shewanella oneidensis*. This link between music, bacteria growth and some orbits is because of the underlying beauty of mathematics (in this case geometric progression).

There was more (of which, a bit more anon), but I decided the best thing was to send the whole letter to Murch, back in London, for his response. Setting aside the swipe at Lineweaver—as Lineweaver, as we have seen, had in the meantime responded to his critics, and they in turn had fired back, as then in turn in turn had he (that was an ongoing battle)—I decided to ask Murch, for starters, what he had made of that last contention of Duffy's, that Titius-Bode was merely an instance of the sort of geometric progression that occurs all over nature. "I don't get Alan's problem, if any," Murch shot back (a bit peremptorily, it seemed to me, but still). "In fact, he seems to be validating Bode by giving examples of other instances in nature where doublings occur. Along those same lines: why, for instance, does the intensity of light drop off by *inverse* doublings?" But he went on to note that, in any case, he was saying something different—for one thing, most of Duffy's instances were temporal, whereas his was spatial—in turn, adding something fresh to such traditional double compoundings: "The particular kink of T-B-M is that the orbital doublings start *at a certain distance* (*β*) from the central object (the Sun or Jupiter or whatever) (what astronomers call *the primary*)," and that since that was not the sort of thing one saw in your everyday geometrical progressions, something else had to be going on as well.

But what about the alternative explanation, the one about orbital resonances that was a refrain I had heard from several of my other interlocutors—thus, for instance, from Greenberg, who had written

> One pattern that comes up a lot, and that does have physical meaning, is the commensurability (small whole-number ratios) of orbital periods. These commensurabilities are common in our solar system and have been found to also be common among extra-solar planets. They are meaningful because there is a physical process that explains them. The small-whole-number ratios mean that geometrical configurations among the bodies are periodic, which enhances their mutual gravitational effects, by periodically repeating the same forces. The result affects how and where planets and satellites can form, and where they may evolve to. But we can say that because we understand the physics, not because of the numerology.

They all claimed already to understand this phenomenon just fine without any recourse to Titius-Bode, with or without Murch's refinements—and indeed, they understood it in the only terms in which they were finally interested, as a product of physical processes playing out at the very outset of the birth of the solar system, when swirling gasses across all sorts of highly sophisticated mathematics congealed by way of such *dynamic* (as opposed to static) processes as orbital resonance. (Thus, granting that Murch might indeed be on to something with his Bodean alignments, though still insisting they likely grew out of orbital resonances, Duffy went on to surmise that "While the exact method by which these approximate regions are populated is unknown, it is almost certainly a complex interaction within the gas cloud from which our Sun (and planets/moons) formed under Newtonian gravity.")

In response to such criticisms, Murch would usually begin by saying that Titius-Bode, far from being a static imposition, was itself quite dynamic in its ongoing unfurling—witness how it seemed to deal with such relatively recently captured objects as Neptune's most distant and elliptically orbiting moon, Nereid, or with several of the outer moons of Uranus and the other outer planets, gradually slotting them into their respective "troughs" (or else, one of those uncannily midpoint renegade orbits) long after the original formation of their various orbital systems. As for the dynamics of origin, Murch *had* hazarded such guesses as the standing wave; his astrophysicist interlocutors just refused to entertain them. But, in this instance, he went on to further point out that orbital resonance in fact occurred much less frequently in our solar system than virtually Bodean orbits. Thus, to take Duffy's appropriation of his own example, it was true that the three inner moons of Jupiter (Io, Europa, and Ganymede) did evince that 1:2:4 orbital resonance (the innermost one, Io, orbiting four times for every two times of Europa and every one of Ganymede), but the fourth moon, Callisto, exhibited no such tendency at all; beyond that, it was the three *outer* moons (Callisto, Ganymede, and Europa) that slotted into Bodean orbits, while in this instance, it proved the innermost one, Io, that was the outlier. In the case of Io, it may in fact be that renegadery may have been reinforced by Io's orbital time being close to a 2:1 resonance with Europa, the next moon out. (Another renegade, Neptune, is similarly locked in a 2:3 temporal resonance with Pluto.) But resonance does not seem to occur with the other three midpoint-slotted renegade moons: Titan, Oberon, and Galatea.

More important, though, Murch went on to note that

The essential difference between Bode and Resonance is that with Resonance the masses of the objects are a factor, alongside their temporal relationship. For this reason, resonance can both attract (into a stable orbital relationship i.e. Io and Europa) but also eject (gaps in the Asteroid belt, where the 2:1 resonant relationship with Jupiter causes asteroids to be ejected from their orbit). The difference is that Io and Europa are roughly the same mass, whereas Jupiter and the average asteroid are eight orders of magnitude (100 million times!) different in mass.

But with Bode, the mass of the planet does not enter into the equation. Ceres fits Bode as accurately as Jupiter (seven *orders of magnitude* difference in mass). This is confounding, because it hints at some other force (ie dark matter or ???) being the regulator. I mention dark matter just as a "handle" to fumble with the concept. Whether it is dark matter or something else, it is in any case "dark" in the sense that we do not know much if anything about it.

"It seems," he concluded,

> as if the interaction of Bode and Resonance works something along the lines of: Bode gets things roughly in place and then resonance, if it happens to exist in any specific case (ie, Io and Europa), will tweak the relationship further. In the case of Io, its resonance with Europa appears to have "tweaked" it right into that renegade orbit!

AT ONE POINT, Duffy suggested that "What Walter should try to do (and which would be highly celebrated) is if he could show, using the existing tools of astronomy, how Newton's gravity and planetary/moon formation from gas disks could necessarily lead sometimes to orbital resonances between planets and hence ratios between their orbital sizes that follow the pattern seen in T-B," and went on to suggest that this would likely require "advanced computer simulations."

Others suggested that all Murch's observations were meaningless absent an extremely technical statistical analysis of the kinds of numbers he was observing.

Greenberg, for his part early on, had been even blunter:

> Great power-point (or video) presentations are not necessarily an indication of meaningful science. Scientific results are presented in refereed papers. I suggest your friend read a few as models (and make sure he understands them in detail), and then try writing up his results. Writing a good

paper is about 75 percent of the research process. Writing is hard work, and forces you to get it right, as you know. A Powerpoint presentation alone is meaningless.

In other words, unless Murch could speak to astrophysicists in their own language, by way of their own channels and methodologies, his ideas would almost by definition be suspect and carry no standing. And by their language, their channels, their methodology, they meant this sort of thing (from one of the Lineweaver/Bovaird papers):

the form:

$$f(R) = \frac{df}{d\log R} = \begin{cases} k(\log R)^{\alpha}, & R \geqslant 2.8\,R_{\oplus} \\ k(\log 2.8)^{\alpha}, & R < 2.8\,R_{\oplus} \end{cases} \tag{5}$$

where $k = 2.9$ and $\alpha = -1.92$ (Howard et al. 2012). The discontinuous distribution accounts for the approximately flat number of planets per star in logarithmic planetary radius bins for $R \lesssim 2.8\,R_{\oplus}$ (Dong & Zhu 2013; Fressin et al. 2013; Petigura et al. 2013b; Silburt et al. 2014). For $R \lesssim 1.0\,R_{\oplus}$ the distribution is poorly constrained. For this paper, we extend the flat distribution in log R down to a minimum radius $R_{low} = 0.3\,R_{\oplus}$. It is important to note that for the Solar System, the poorly constrained part of the planetary radius distribution contains 50 per cent of the planet population. For reference the radius of Ceres, a "planet" predicted by the TB relation applied to our Solar System has a radius $R_{Ceres} = 476\,km = 0.07\,R_{\oplus}$.

The probability that the hypothetical planet has a radius that exceeds the SNR detection threshold is then given by

$$P_{SNR} = \frac{\int_{R_{min}}^{R_{max}} f(R)dR}{\int_{R_{low}}^{R_{max}} f(R)dR} \tag{6}$$

We do not integrate beyond R_{max} since we expect a planet with a radius greater than R_{max} would have already been detected. We define R_{max} by.

$$R_{max} = R_{min\,SNR} \left(\frac{P_{predict}}{P_{min\,SNR}} \right)^{1/4}, \tag{7}$$

where $R_{min\,SNR}$ and $P_{min\,SNR}$ are the radius and period respectively of the detected planet with the lowest signal-to-noise in the system. $P_{predict}$ is the period of the predicted planet. R_{min} depends on the SNR in the following way:

$$R_{min} = R_{\ast} \sqrt{SNR_{th}\,CDPP} \left(\frac{3\,hrs}{n_{tr}\,t_{T}} \right)^{1/4}, \tag{8}$$

where SNR_{th} is the SNR threshold for a planet detection, n_{tr} is the number of expected transits at the given period and t_{T} is the

The way such things are supposed to be presented

George Bernard Shaw once famously quipped that every profession is a conspiracy against the laity, but the priesthood of advanced science nowadays is especially well armored, in particular thanks to its mandarin reliance on mathematics that is well-nigh impenetrable to average and even well-above-average followers. This was not always the case. As Murch himself has pointed out, thanks to the scandal of "unspeakable number *arrhetos*," the dread Pythagorean discovery of the insoluble irrationality of the square root of two ("a horrible flaw that was only revealed to the Pythagorean student once he'd ascended to a sort of brown belt level, and even then only on condition that he never breathe a word of it to anyone else"), the natural scientific tradition growing out of Aristotle (including the quadrivium as a whole, and the music of the spheres in particular), while steeped in number, relied on little more than arithmetic and geometry. The remathematization of the sciences began again with Galileo, who insisted that "Nature writes her book in the lan-

Faraday

guage of mathematics" and has been compounding exponentially ever since—though not to everyone's delight. Michael Faraday, arguably one of the greatest scientists of all time though mathematically virtually illiterate, used to complain that "Facts have preceded the math, or where they have not the facts have remained unsuspected though the calculations were ready...and sometimes when the fact was present, the calculations were insufficient to illustrate its true nature until other facts came in to help."

Yet the tendency in physics over the past century has been toward greater and greater mathematical abstraction (culminating in the apotheosis and current hegemony of string theory), "a development," according to the lapsed physicist and celebrated science writer Margaret Wertheim, "that seems to both capture nature precisely and at the same time erase it. None of us," she goes on

Wertheim

will ever see the four dimensional spacetime of general relativity or touch the eleven dimensional manifold of string theory. These extraordinary concepts describe a concept that shimmers like a mirage, exquisite and intangible and increasingly unreachable by the human body and mind. I hold degrees in physics and mathematics and writing about these subjects is my profession, yet I struggle to understand general relativity. Any nonphysicist who says otherwise is not being honest.

Wertheim goes on to suggest that this—not supersymmetry or multiverses or the precise number of dimensions wrapped into any putative string matrix—is *the* cosmological problem of our day. "Traditionally," she concludes, "the purpose of cosmology was to embed a people in a world—but what happens to a society when its official cosmology becomes one that 99 percent of its population does not understand and very likely can never hope to understand?" At which point, she goes on to suggest that we may be approaching a moment calling out for an equivalent of the Protestant Reformation, when Luther insisted that a sacred text heretofore only available to a priesthood versed in Latin or Aramaic or Hebrew be made available in vernacular languages so that ordinary people would be able to read it for themselves.

Curiously, for his part, Murch is not all that bothered by that hypermathematization; indeed, he'd like to take part. He'd love to work with any physicist who would have him on a rigorous statistical analysis of the tolerances of his ratios, or on sophisticated computer modelings and the like: he just can't find a single physicist willing to take him on. Margaret Wertheim, whose exceptionally lucid and entertaining book on the impasse between conventional scientists and so-called outsider physicists, *Physics on the Fringe* (2011), was my source for the quotations just given, told me in conversation recently, "What you and Walter have to understand is, that's the universal complaint of *all* outsider physicists: 'Why won't they just engage with us?'" She can relate: her own book, widely praised in general and art world venues, has yet to receive a single notice in any scientific publication.

But why is that? I asked her (not so much about the reception of her book as the reception of outsider ideas among the astrophysical profession in general). "Well," she began, "for starters, there's the problem of cranks. You can't imagine how many such screeds these professional people get, individuals convinced that they and they alone have cracked the secret of the universe. And when the outsiders demand, 'Why won't you just give us a hearing?' the scientists are often justified in answering, '*Because you just don't listen!*'" (Although Wertheim had made a point of saying that Murch was a special case, clearly much more sophisticated in his background and his analysis than the average outsider, I wondered to what extent this criticism might apply to him as well. He, too, always seemed to have a comeback, the difference perhaps being that it was seldom a mere repetition of the opening gambit and more usually a fascinatingly pertinent elaboration of the theory.)

Wertheim's surmise reminded me of a piece in the October 2015 issue of *Scientific American* wherein their Skeptic columnist Michael Shermer described attending the recent Electric Universe (EU) conference in Phoenix, Arizona—a veritable convocation of inspired outsiders, one of whose members characterized Einstein's general theory of relativity as basest "numerology," and another of whom patiently explained to Shermer the impossibility of peer reviewing their work: "In an interdisciplinary science like Electric Universe, you could say we have no peers, so peer review is not available." When Shermer asked the conference host if EU theory offered anything like the practical applications that theoretical physics had given the world—GPS telemetry, the pinpoint ability to land spacecraft on hurtling comets, and the like—the convener blithely responded no. Then what, asked Shermer, does EU theory add? "A deeper understanding of nature," he was told, to which Shermer responded, "Oh." (Though here, too, I found myself wondering to what extent Murch would be immune to the same response, and to what extent the demand for immediate practical outcomes might have eluded many of the great natural historians of the nineteenth century, starting, say, with Darwin.)

But, again, I persisted with Wertheim, Murch is hardly your average outsider: he is every bit as accomplished in his field of sound design (and the subfield of acoustics generally) as any of the astrophysicists he's been trying to engage are in theirs. Why couldn't he parlay any sort of full engagement? "There's also the question of time," Wertheim responded. "You can't imagine how busy these people are, how swamped with committee work and budget reviews and supervision of underlings, not to speak of the requirements of their own work." (Actually, I could. It had been a constant refrain in all my own attempted contacts. As Duffy said, in apology, all too typically, at the outset of his communiqué, "Sadly it's often the case that opportunities to discuss ideas from researchers outside the establishment can be time-consuming and—from the researchers' perspective—unproductive as they are usually obviously flawed from the start. In a world where we have little time to perform our own research—I typically have to do mine in the evenings and weekends—it's impossible to find time to discuss in great detail someone else's.") Then, too, as Wertheim pointed out, there was the problem of the ever-tightening sense of specialization, such that professional astrophysicists either actually felt or at least could use the excuse of feeling incompetent to discuss matters outside their ever-more-narrowing silos of expertise.

The wider problem, though, concerned the more general profile of the sociology of academic science in the current era and the pressures within it that mitigated against entertaining any sorts of ideas that failed to fall squarely in the mainstream (especially ideas as notoriously cootied-up as Titius-Bode). In a chapter in his book *The Utopia of Rules: On Technology, Stupidity, and the Secret Joys of Bureaucracy* London School of Economics anthropologist David Graeber (one of the founders of the Occupy Wall Street movement) wonders why we have all been denied the flying cars and transporter beams and so forth that we were promised in the futuristic prognostications of the fifties and sixties. "There was a time," he hazards by way of answer, "when academia was society's refuge for the eccentric, brilliant and impractical. No longer. It is now the domain of professional self-marketers. As for the eccentric, brilliant and impractical: it would seem that society has no place for them at all." Graeber goes on to quote the physicist Jonathan Katz's warning to students pondering a career in the sciences, how even when one does emerge from the usual decades-long period languishing as someone else's flunky, one can expect one's best ideas to be stymied at every point, since "It is proverbial that original ideas are the kiss of death, because they have not yet been proven to work." The sorts of thinkers "most likely to come up with new conceptual breakthroughs," Graeber concludes, "are the least likely to receive funding, and if somehow a breakthrough occurs, it will almost certainly never find anyone willing to follow up on the most daring implications."

Graeber's relatively cursory analysis was explored in considerably more damning depth in Lee Smolin's 2006 book,

Smolin

The Trouble with Physics. "A hundred years ago," notes Smolin, a founding member of the innovative Perimeter Institute for Theoretical Physics in Waterloo, Ontario, "the academy was much smaller and much less professional, and well trained outsiders were common. This was the legacy of the nineteenth century, when most of the people who did science were enthusiastic amateurs." But things have changed radically, particularly since the 1970s, when universities stopped growing but kept churning out Ph.D.s in physics and the other sciences. "As a result, there is fierce competition for places in research universities and colleges…and also much more emphasis on hiring faculty who will be funded by the research agencies. This greatly narrows the option for people who want to pursue their own research programs…and there are fewer corners where a creative person can hide, secure in some kind of academic job, and pursue risky and original ideas." On top of that, recent years have seen "a marked increase in the number and power of administration" such that "in hiring there is less reliance on the judgment of individual professors and more on statistical measures of achievement, such as funding and citation levels," which "also makes it harder for young scientists to buck the mainstream." Smolin also goes on to note how "peer review" serves a powerful gatekeeping function, "a mechanism for older scientists to enforce direction on younger scientists" and "to discourage change."

Put simply, Smolin continues, "the physics community is structured in such a way that large research programs that promote themselves aggressively have an advantage over smaller programs that make more cautious claims...To do the opposite—to think deeply and independently and to try to formulate one's own ideas—is a poor strategy for success" (and may, for example, account for why someone like Lineweaver, pursuing independent research on as contested a terrain as Titius-Bode, might still be only an associate professor after fifteen years, as Falco so archly noted; or, for that matter, why Murch was probably never going to find a grad student willing to risk his own career by working with him at any great length). Smolin insists that in order to prosper, science requires both rebels and conservatives, both seers and craftspeople; the current state of play, however, mitigates mightily against the professional survival of rebels and seers. No wonder, as he contends, theoretical physics has effectively been stuck for more than thirty-five years, ever since the consolidation of the so-called standard model and the field's massive detour into string theory.

Smolin himself delved deeply into string theory years ago, but he began to lose faith in the exponentially complexifying field as he became convinced that no matter how elegant and beautiful the theorizing might grow (albeit in its achingly refined way), it was becoming less and less likely that any of it could ever be proven (or even falsified) one way or the other. A large part of his book becomes, for starters, a sort of lay exposition of the contours of this self-reinforcing and now utterly dominant way of thinking (at one point, he notes how "in the country's top physics departments—Berkeley, Caltech, Harvard, MIT, Princeton and Stanford—twenty out of twenty-two tenured professors who received their Ph.D.'s after 1981 made their reputations in string theory or related approaches"), which he follows with a blistering critique of the movement's pretensions.

Margaret Wertheim made similar points in her book, and in conversation with me: at one point she described having gone to two conferences, one hard on the other, a convocation of outsider physicists and then a colloquium of professional theorists celebrating the new field of string cosmology, and how she had found it hard to figure out which was the loopier, or the less tethered to any sort of reality check. If anything, she found the latter more wildly "surreal." The outsider conference featured one hundred twenty theories on parade, the string cosmology meeting seriously broached the possibility of 10^{500} (that's 10 followed by 500 zeros) legitimate variants—there are, Wertheim noted dryly, only 10^{80} subatomic particles in the entire known universe. "Utterly splendid," she quoted one rapt string enthusiast at the latter conference as exulting, whereupon he continued, "Of course, not a shred of evidence for any of it." Whether there were 10 or 11, or 26 dimensions—who knew?—they were all wrapped tight one inside the other at scales physically impossible to gauge. ("Like how many angels you could fit on the head of a pin," one was tempted to say, only it's worth acknowledging how, back in medieval days, if one actually believed in the existence of angels, such would have been an entirely legitimate and indeed urgent question.) How come, I asked Wertheim, those professional physicists are able to stake entire careers on the existence of such infinitely infinitesimal, and almost by definition unverifiable, dimensions when Murch gets attacked for suggesting the existence of his heretofore unmeasured but in principle not unmeasurable sorts of waves? That, she said, was a good question.

Surveying the sheer hegemony of string theorists among most academic departments today, Smolin notes how "It turns out that sociologists have no problem recognizing this phenomenon. It afflicts communities of highly credentialed experts who by choice or circumstance communicate only among themselves...and because its consequences have sometimes been tragic, there is literature describing the phenomenon, which is called *groupthink*." Smolin acknowledges that in response to charges of this sort, string theorists regularly reply that the opinions of outsiders should be disregarded because those outsiders are simply not skilled enough to evaluate the evidence or alternatively (the charge leveled against someone like Murch) advance any of their own.

Smolin argues that the only way out of this sort of impasse is to put forward a theory of how science ought to work. He begins by noting that human beings "are masters at drawing conclusions from incomplete information...But this ability comes at a heavy price, which is that we easily fool ourselves"—this being Smolin's way of talking about what Murch calls apophenia. "I believe," Smolin goes on, "that science...is a way to nurture and encourage the discovery of new knowledge, but more than anything else, it is a collection of crafts and practices that, over time, have been shown to be effective in unmasking error. It is our best tool in the constant struggle to overcome our built-in tendency to fool ourselves." Leaning heavily on his own mentor, the philosopher Paul Feyerabend, Smolin goes on to argue, however, that "even in cases where there is a widely accepted theory that agrees with all the evidence, it is still necessary to invent competing theories in order for science to progress," for as Feyerabend noted (in Smolin's paraphrase), "When scientists come to agreement too soon, before they are compelled to by the evidence, science is in danger...Science moves forward [by contrast] when we are forced to agree with something unexpected. If we think we know the answer, we will try to make every result fit that preconceived idea"—or, as I myself might add, prejudice—"It is controversy that keeps science alive."

The philosopher of science P. Kyle Stanford at the University of California at Irvine went even further along these same lines in his 2006 book, *Exceeding Our Grasp: Science, History, and the Problem of Unconceived Alternatives* (a volume in turn recommended to me by Stuart Firestein, the chair of biology at Columbia University, and the author himself of books by turns celebrating, as their titles suggest, the roles of *Ignorance* and *Failure* in the ongoing advance of science). Stanford's book is concerned, as he writes near the outset, with "the possible existence of alternatives to our best scientific theories that share some or all of their empirical implications—that is, quite different accounts of entities and/or processes inhabiting some inaccessible domain of nature that nonetheless make the same confirmed predictions about what we should expect to find in the world... that our own theories do." He goes on to cite a whole series of historical examples that as a group suggest "a robust, distinctive pattern in which available evidence cited in support of each earlier theory turned out to support one or more competitors unimagined at the time just as well." For that historical claim to hold up, he goes on, "we must explicitly note that a theory need not explain or accommodate all the existing data in order to count as well confirmed; evidential anomalies are allowed. The point is that we have repeatedly been able to conceive of only a single theory [at a time] that was well supported by all the available evidence when there were indeed alternative possibilities, *also* well supported, indeed perhaps equally or even better supported by the same body of evidence." Stanford goes on to propose at length an alternative way of thinking about theories—that none of them should be conceived as literally true of the world out there; that all of them should be entertained as provisional approximations, working guesses (we are once again in Vaihingerian terrain, though, surprisingly, Stanford never mentions him); "and that the fundamental theories of contemporary science should be regarded, like their historical predecessors, simply as powerful conceptual tools for action and guides to further inquiry rather than accurate description of how things stand in otherwise inaccessible domains of nature," and concludes that the failure to do so, and the wholesale rejection of alternative theories, has repeatedly held back the progress of vital science. (See, for example, as Murch himself likes to point out, as well, the way in which the wholesale vilification of Jean-Baptiste Lamarck through much of the twentieth century forestalled the recognition of all sorts of insights about evolutionary succession and the heritability of acquired characteristics that are now making their way into acceptance, through the back door, as it were, by way of breakthroughs in the field of epigenetics.)

All of this of course struck me as obviously pertinent to Murch's situation, and I was especially heartened to see how in laying out his principles for a just and future science (beginning "We agree to argue rationally, and in good faith, from shared evidence, to whatever degree of shared conclusions are warranted"), Smolin for his part explicitly insists that "membership in the community of science [be] open to any human being." I was therefore somewhat disappointed, when I contacted Smolin and asked him to review Murch's video and writings, that his response pretty much tracked with most of the others (pleading lack of time though discounting the "gravitational waves" hypothesis pretty much out of hand). Going back to his book, however, I noticed that he elaborates on his "open to all human beings" edict with a proviso that "entry to the community of science" be limited to people who have attained "mastery of at least one of the crafts of a scientific subfield to the point where you can independently produce work judged by other members to be of high quality" (which of course brings us back to the problem of Murch's inability to frame his arguments in the language and using the protocols of peer-review publication and statistical analysis and the like upon which conventional scientists insist, and then the catch-22 of the fact that precisely because he can't, none of them seems willing to assist him in doing so).

It is possible to imagine an alternative way of proceeding in the Republic of the Imagination, one suggested, for example, by the artist Robert Irwin's conception of the dialogue of immanence, as he calls it, in which the leading practitioners in all sorts of disciplines who are pushing out toward the very limit of their respective fields (art, music, architecture, technology, information science, physics, biology, cosmology, and so forth), who indeed are themselves, by way of their engagement in the pursuit of what he calls "pure inquiry" extending the very boundaries of those disciplines—how those people keep bumping up against one another and ought to have much to say to one another, if only they are willing to listen and engage. (In this context, mightn't it be possible that a sound designer as accomplished as Murch could have, if nothing else, fresh metaphors regarding such things as standing waves to offer astrophysicists intent on their own projects?) Einstein for his part always insisted that in the pursuit of truth, imagination was far more important than technical knowledge. Vladimir Nabokov, the great lepidopterist, suggested (counterintuitively) that "the true master needs the precision of the poet and the imagination of the scientist."

Irwin   *Nabokov*   *Foucault*

And Michel Foucault confessed, in his essay "The Masked Philosopher" (in a passage that serves as the epigraph to Wertheim's book), how:

> I dream of a new age of curiosity. We have the technical means for it; the desire is there; the things to be known are infinite; the people who can employ themselves at this task exist. Why do we suffer? From too little: from channels that are too narrow, skimpy, quasi-monopolistic, insufficient. There is no point in adopting a protectionist attitude, to prevent "bad" information from invading and suffocating the "good." Rather, we must multiply the paths and the possibilities of coming and goings.

Or, as John Cage once said:

> I can't understand why people are frightened of new ideas. I'm frightened of the old ones.

AS I SAY, Walter Murch just knows stuff.

He knows that Columbia University was built on the grounds of the former Bloomingdale Insane Asylum and that because, back in 1892, it was boldly decided to electrify the whole campus and the surrounding Morningside Heights neighborhood, deploying Thomas Edison's then-current DC current, part of the Upper West Side remained in a sort of cultural-technological isolation, long after the rest of the country had adopted Nicola Tesla's rival (and much more efficient) AC current, such that Morningsiders had to rely on wind-up record players and nonelectrical iceboxes filled with regularly delivered blocks of real ice well into the 1940s. (He knows this because he lived it.) He knows that the three fathers of cinema were Beethoven, Flaubert, and Lumière (that's not so much knowledge as a theory of his, which he is happy to expound upon at length). He knows that, at the cellular level, vision is occasioned by interruption of the neurotransmission of glutamate, which of course leads to the (to him) obvious question "What happens to people who are losing their sight, and this neurotransmitter is no longer interrupted but instead streaming constantly?"— which naturally in turn leads to a discussion of "the delightful (unless you're experiencing them)" hallucinations of Charles Bonnet Syndrome. "That Charles Bonnet being the same one into the translation of whose *Contemplation de la Nature* our friend Johannes Dietz (Titius) originally feathered his curiously profligate algorithm." Murch knows that Peter Medawar, the Nobel Prize–winning biologist, declared in his 1959 Reith lectures that "Our scientists' complacency can be traced, I suppose, to an understandable fault in temperament: scientists tend not to ask themselves questions until they can see the rudiments of an answer in their minds. Embarrassing questions tend to remain unasked, or if asked, to be answered rudely." (He knows this because he read it just last night in Arthur Koestler's 1967 book, *The Ghost in the Machine*, his current bedside reading, and hastened to e-mail it to me.)

Murch knows that the structure of the brain (neurons and synapses two millimeters across—Université Laval) looks like the image on left on the next page; and that the structure of the universe (one billion light-years across, with

each bright pixel being a galaxy—Max Planck Institutes)
looks like the image at right;

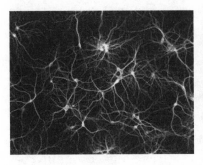

Microstructure of the brain   *Macrostructure of the universe*

and that the Cambridge mathematician John Barrow once
said, "Any universe simple enough to be completely understood would be too simple to produce a mind capable of
understanding it."

Notwithstanding which, Walter Murch also knows, as
he once wrote me, larkishly, that

> it takes a million years for a highly energetic gamma-ray
> photon to travel from the center of the Sun (where it is
> generated by fusion) to the edge of the Sun. By the time
> it reaches the edge, it has used up so much energy collid-
> ing and being collided with that it is now just a friendly
> "yellow light photon." Once freed from the embrace of
> the Sun, this yellow-light photon takes only eight minutes
> to travel to the Earth, where it hits your eye (should you
> be looking at the Sun, which is not advised). The inside
> of the sun is completely dark and contains no atoms: It
> is just a hot soup (plasma) of free electrons, photons,

protons, and neutrons, so hot that electrons, should they try to "marry up" with protons, are immediately stripped away by the force of all the collisions going on. The Darkness of the Interior of the Sun comes from the absence of these electrically neutral atoms. Everything (free electrons and protons) is electrically charged in the Sun (atoms—paired negative electrons and positive protons—are transparent to light and allow it to pass). Inside the Sun, the situation is like some kind of daemonic bumper-car ride at 3am in the Palisades Amusement Park of Hell, a situation we should be happy about, because were it not for this million years of bumper dodge-em cars in the dark, the energy of the gamma-ray photons would not be depleted (by a factor of 2,600), and they would consequently come streaming out of the sun, annihilating with radiation any complex molecular substance (like you or me) that got in their way.

Murch knows that Venus's orbit is very circular and so is Neptune's, which is the most circular of any of our sun's planets, which is odd, since it is usually the outer planets that feature the more elliptical orbits (and indeed Pluto's is the most elliptical). He knows that Uranus rotates on its side, which is weird, but not as weird as Venus, which rotates *upside down* (it is the only planet in our system that rotates counterclockwise). He knows that Mercury's orbit, which is surprisingly elliptical, is relatively far from the Sun, especially in comparison with Neptune's innermost moons, which, relative to the volume of their respective primary, orbit much, much tighter in.

Wait. Whoa! What? *What!?* Isn't beta (β), the limit of negative infinity in Murch's Bodean variation, supposed to constitute the innermost trough in Murch's wave system and hence be consistent from primary to primary? (How could there be orbits in tighter, relatively speaking, than Mercury's, which is already at negative infinity?) Not at all, Murch corrects me. Beta is different for each primary (each star, sun, and planet—every gravitational center), and it's just that once you establish beta, from there on it is doublings (at first almost infinitesimal, and not necessarily occupied) one after the next all the way out.

Well, then, I ask Murch by e-mail, how do you establish beta for any given orbital system? In answer to which I get a digital mouthful. "There are two ways to establish β. One simple, called 'curve fitting'; the other more complex but potentially more interesting, derived from the physics of the system in question. It would be great to consolidate both together, but some of the detailed math of the second system is beyond me." The first, curve fitting, turns out to be hardly as simple as all that, but the gist, as Murch suggests, is that "If a system (like HR8799, or Jupiter's moons, or the Solar System, or...) has planets that show Bodean distance ratios (relatively speaking) such as Ceres/Mars (1.78) or Earth/Venus (1.42) or Mars/Earth (1.6) etc., you can work backwards to the origin (Bode value 1) and then fit in the actual measurements (in kilometers) of that system and get the value of β, which occurs when Bode has a value of 1." And indeed, that will be a different place relative to the primary for each different orbital system.

Murch's other way of deriving beta is indeed quite intriguing and "involves comparing the ratio of Rotational Momentum of the Primary (ie the Sun or planet in question) to the Angular Momentum of all the Secondaries (its planets and/or moons). The Sun rotates exceptionally slowly (once every 25 days) and the planets are very far out compared to the diameter of the Sun. This ratio of low rotational momentum to high angular momentum would yield a high value for β. In other words, the diameter of the Sun is a low proportion of its β." Since Neptune, as big as it is, is rotating astonishingly quickly (once every sixteen hours!), and its moons are much tighter in (in fact, with one of them, Triton, orbiting in bizarrely retrograde against-the-flow-of-traffic fashion, indeed uniquely so among all the large moons in the solar system), the Neptunian system has high rotational momentum and low angular momentum, made even lower by Triton's contrary revolution, and this yields a low value for β. In other words, the diameter of the planet Neptune is a high proportion of its β, the opposite of the Sun. Think of an ice skater, Murch suggests helpfully. What happens as she pulls in or pushes out her arms? "With Jupiter, Saturn, and Uranus, the ratios of RM/AM are somewhere moderately in between the Sun and

An instance of angular momentum

Neptune, and consequently, their values for β are moderately in between as well. Their β's are low single-figure multiples of their diameters. All three are similar, and I would say that this is 'normal' whereas the Sun and Neptune are 'abnormal,'" but again, he says, he would need help in working out the math, and it would be nice if he could get some.

It's interesting about Mercury, though, he goes on to muse, because Mercury's orbit is unusually elliptical for an object as close in to its primary. "It's almost as if Mercury is a late addition, a captured object that came barreling in from further out in the solar system, maybe flipping Venus in the process, and that makes the fact of its congruence with the predicted Bode orbit all the more fascinating."

EVER SINCE READING his *Trouble with Physics*, I had been corresponding with Lee Smolin of the Perimeter Institute in Toronto, and fairly late in the process, I decided to send him a working version of this manuscript, to which he presently responded at some length. Starting out by noting that Murch seemed a fascinating individual and one whom he would be happy to talk with at length, Smolin went on to lodge a variety of objections from the now-common (gravitational waves just too weak to be registering these sorts of effects) to the more novel (if there were such troughs and waves, surely someone would by now have noticed anomalies in the trajectories of the thousands of space probes with their extremely exacting systems telemetries), before concluding:

> My sense is that Walter has done just about as well as any lay person could do, without the tools and discipline instilled

by succeeding in a Ph.D. program. The main thing a professional life in science teaches is that almost every idea is wrong. All of us who work in science have seen most or all of our cherished, beautiful ideas fail. This is the tragic element of a life in science. Few are immune. What a life in science teaches one is that science is really, really, really hard to do right.

The main mistake Walter is making is to stay in love with his first idea. Only when you have seen your first 20 or 50 ideas die do you begin to appreciate what it means to have a good scientific idea. This is why laypeople like Walter never succeed in contributing to science—they cannot give up their first ideas.

The second mistake Walter is making is to underestimate the likelihood that a flexible theory can be invented and adjusted to fit random data, even when it is based on wrong ideas. This, unfortunately, happens all the time in medical science. If one has 100 factors to try to correlate with a disease, and one requires 95% probability for a match to be taken as a result, then at least five of them will appear to fit, just by chance. The literature of science and medicine is full of ideas that had some success at a 95% or even 99% level, but on further examination were found to be wrong.

Every year we see "discoveries" of new particles in experiments at the 95% or 99% level, which is to say better than Bode's law does, which go away when larger data sets are taken. This is why the standard for discovery in particle physics is five sigma (one chance in 3 million of arising by chance). Even so, last year there was a five sigma "discovery" in cosmology that got lots of attention before it was shown that the effect could be explained by reflection off of dust in our own galaxy.

Bode's law has if I understand right three free parameters. There were initially six known planets. The right question to ask is how likely is it someone could have invented a rule with three free parameters that fit six numbers. The answer is that given that there are a vast number of simple patterns with three parameters, the probability that something like Bode's law could have been found to match random planetary orbits is close to one. And given Neptune, and the 50% success rate with some systems and the 5% success rate with other systems, the law has done about as poorly as could be expected were it an accidental fit, fine tuned to a small data set.

When a hypothesis works in half the new cases, or when it has a five percent success rate against a new data set, the right conclusion to draw is that the hypothesis is wrong. We must throw such ideas away if science is to be a source of reliable knowledge, and progress.

Later the same day that I received Smolin's note, another friend of mine, John Hastings, a primary school teacher with whom I had been discussing Murch and the response to his ideas, sent me the following uncannily similar quotation: "[His] is not scientific [procedure] but takes the course of an initial idea, a selective search through the literature for corroborative evidence, ignoring most of the facts that are opposed to the idea, and ending in a state of auto-intoxication in which the subjective idea comes to

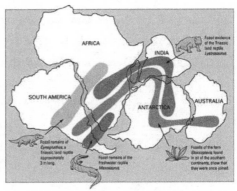

Wegener   *Continental drift*

be considered an objective fact"—an evaluation that turned out to be but a single instance of the overwhelmingly negative contemporary response (in this instance, among professional geologists) to Alfred Wegener's original postulation of his theory of continental drift, back in 1915. When I asked my friend where he had found it, he said he was in the midst of reading Susan Wise Bauer's *The Story of Western Science*, and that Bauer for her part had gone on to gloss the passage as follows: "The wholesale resistance to Wegener's intuitive leap may have had something to do with border protection; Wegener was neither a geologist nor a paleontologist. He was a tinkerer in meteorology, an adventurer who had once been forced to eat his own ponies to survive in an icy Greenland camp, a German who had fought on the side of the Central Powers in World War I."

Indeed, when I went on to research the Wegener story, I found all sorts of other similar attacks, including H. F. Reid's:

There have been many attempts to deduce the characteristics of the earth from a hypothesis: but they have
all failed…This is another of the same type. Science has
developed by painstaking comparison of observations
and, through close induction, by taking one short step
backward to their cause; not by first guessing at the cause
and then deducing the phenomena.

And R. T. Chamberlain's:

Wegener's hypothesis in general is of the foot-loose type,
in that it takes considerable liberty with our globe, and is
less bound by restrictions or tied down by awkward, ugly
facts than most of its rival theories. Its appeal seems to
lie in the fact that it plays a game in which there are few
restrictive rules and no sharply drawn code of conduct.

The geologist Barry Willis, the author of the first quotation
my friend Hastings had sent me, went on to insist that "Further
discussion of [Wegener's ideas] merely encumbers the literature and befogs the mind of fellow students." The afore-
mentioned Chamberlain, for his part, went on to vesuviate,
perhaps unintentionally revealingly, that "If we are to believe
in Wegener's hypothesis we must forget everything which has
been learned in the past 70 years and start all over again."

Intriguingly, Wegener's entire argument at the time was
based on a series of observations, how patterns of mineral
seams or fossil deposits along one coastline rhymed almost
exactly with those found along another coast far distant,
and how coal deposits (resulting from onetime tropical

forests) could be found in the Arctic, and other evidence suggested onetime glaciation in the sub-Saharan tropics. He amassed dozens of these observations but was the first to acknowledge that he still lacked a comprehensive theory as to how such drifting of continents could ever have come to be. He hazarded two guesses—one based on the centrifugal force of the rotation of the earth, and the other, a so-called tidal argument, based on the gravitational attraction of the Sun and Moon—both of which were dismissed by his professional critics with howling scorn because the supposed mechanisms were (in a familiar formulation) "orders of magnitude too weak." Wegener, however, was not deterred, acknowledging that "The Newton of drift theory has not yet appeared. His absence need cause no anxiety; the theory is still young." Indeed, that Newton, those Newtons, did not arrive until the early 1960s, when, one after another, a series of papers (by Robert Dietz in 1961—that same name again, if this were the eighteenth century it would be rendered *Titius!*—and Harry Hess in 1962, among others, both drawing on the at-the-time largely dismissed and unsung work from the previous decade by a Columbia University oceanographer named Marie Tharp, who played much the same role in this discovery as Rosalind Franklin did in the roughly contemporaneous discovery of DNA), determined that such continental drift could indeed have resulted, and almost certainly did so, by way of the newly discovered mechanisms of mantle convection currents ("seafloor spreading") and plate tectonics.

I wrote to Murch to see if he had ever heard of Wegener, and naturally he had:

Yes, that theory is a great example of the phenomenon. In seventh grade science, back in 1955, I approached my teacher with what I thought was an original concept: that South America and Africa must once have fit together, and he waved the idea away: "It looks that way, but we know this is not the case. There is no mechanism for continents to move around like that."

Six years later, I went to Johns Hopkins intending to be an Oceanographer, and prerequisites involved taking Geology. I enjoyed it, to a certain extent, and still enjoy looking at some aspects of the world geologically, but something did not click between me and the subject matter, and the following year I switched to Romance Languages and History of Art. Looking back on the class from this distance, I think part of my problem was that continental drift had not yet been accepted in 1961, and so the "story" of geology lacked some essential ingredient— some would say "the truth"—to make it compelling to me. To be uncharitable, it seemed (to paraphrase Ernest Rutherford) a bit like stamp collecting.

Wegener also did not speculate on the causes of drift, as I remember it. No one at that time could figure out the mechanics of the drift. Instead, Wegener just pointed out ample evidence of ancient connections (the identical fossils occur in the "nose" of Brazil as in the "armpit" of Africa).

One always has to be careful how causal speculation is handled. Evidence of a pattern requiring more investigation, and speculation of the causes of that pattern are two different things. Most famous is Newton's reluctance to speculate about the cause of universal gravity and its non-intuitive nature: *Non fingo hypothesi*. How could something influence something across huge distances of empty space? There was no answer to this, even provisionally, until Einstein came along 150 years after Newton.

I am pointing out a statistical pattern, based on a simple formula, that requires (I believe) further investigation. I am also pointing out the interesting fact (never before revealed) that the Bode formula has a family resemblance to the formulae that we encounter in music theory; and that the spacing of the potential orbits can therefore be converted to musical intervals, which has historical resonance with the ancient theory of the Music of the Spheres. Although the two Musics are very different in intervals and in the basis of the calculation of those intervals.

That's really all I am doing. After that, and one does have to be careful how one presents this, officially speaking, *non fingo hypothesi*.

It was funny: now that I was coming to an end of my reporting, everybody had ideas about how I should present the results. Smolin, for his part, ended his generous note as follows:

Even if Bode's law is wrong, or a rough consequence of some messy astrophysics, you have a beautiful story to tell. Just please don't frame it as a heroic story of a lay person who saves science from its corrupt and evil masters. Use the story to explain why science works. This is important. Climate change deniers, creationists and vaccine refusers thrive in that dark space where the ability of science to select reliable knowledge from a vast sea of wrong ideas is not understood.

That last riposte of Smolin's seemed to me the most unsettling of any I had heard, for indeed, how might climate change deniers instrumentalize a story like this? Still, what if notwithstanding all that, Murch were on to something? My primary school teacher friend John Hastings summed up his take: "My favorite is when you're reading something and you agree with every single person as they are talking—and yet at the same time you know they can't all be right."

FROM THE CLOSEST reaches of the solar system (upside-down Venus and uncanny Mercury) to its farthest, Murch now began corresponding with me about the so-called Kuiper asteroid belt beyond Neptune, the millions and millions of ice objects (as opposed to the largely iron and stonelike ones spread out between Mars and Jupiter) that languor out on the farthest edges of our solar system, every once in a while coming dislodged to form comets or else get captured as moons by some of the outer planets. The thing is, there's a big mystery about the Kuiper Belt. The belt seems to start up, fairly abruptly, right around the orbit of Neptune, but then it subsides, incredibly abruptly, just like that, within a few hundred thousand miles—there's stuff, and more stuff, and more stuff, and then, suddenly, apparently, there's none—along a diameter ridge about 50 AUs from the Sun. The mystery of the Kuiper Cliff, as it is known, shows up on all sorts of lists of the solar system's biggest mysteries, in part because the conventional explanation is that there must be some sort of relatively large "shepherding" planet, some so-called Planet X, out there beyond the orbit of Pluto, that is effectively hemming the Kuiper Belt in with its gravitational force. Only, notwithstanding the appearance of the planet in all sorts of science-fiction fantasies (such as, most recently, Lars Von Trier's *Melancholia*), nobody has been able to spot it, and it hasn't been for lack of trying—which at a certain point begins to eat away at that hypothesis.

One day recently, in passing over the phone, Murch let drop that, oh incidentally, his variation on Titius-Bode thoroughly addresses the problem. (*Talk about burying the lead!*) He went on to note that he had had some slides about that late in his keynote lecture carousel, only he hadn't gotten to them in his videotaped talk. He directed me to the slides in question, and indeed they offered, it seemed to me, some of the most dispositive evidence yet in support of his theory.

Because Murch now focused on the situation in the trough between the Bode wave peaks at $n = 6$ and $n = 7$, which is to say between 28 AUs and 50 AUs. Neptune—recall how it was exactly halfway between the two places it was supposed to be—indeed rests near the top of the inner slope on the $n = 6$ (sun-closest) side, at 30.1 AUs, which is, as we have seen, just about where the Kuiper Belt starts. Pluto—remember how it turned out to be exactly where it was supposed to be—is at the very bottom of the trough, at 39.5 AUs, except not exactly. Because Pluto's is an extremely elliptical orbit, indeed at its closest to the sun, it pulls inside the orbit of Neptune, at 29.6 AUs, but then at its farthest, it almost, but not quite, reaches the peak of the wave on the far other side at 49.3 AUs. Haumea, another larger-than-average, recently discovered Kuiper Belt object, a little farther out than Pluto with its mean average orbit at 43.13 AUs, negotiates pretty much the same sort of trajectory, up the inner slope and then down it and way up the outer one, almost reaching the top lip of the peak but not quite—something seems to be blocking it, some insurmountable pressure—and then back down again. And the same holds true for *almost all* the other Kuiper Belt objects: hardly any seem able to push past the peak of the wave at n = 7, the sort of area in the Sun's orbital system that Murch had earlier predicted would be barren.

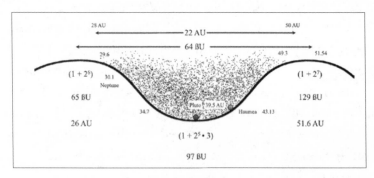

The Kuiper Belt

Murch describes the effect as an incidental one within the context of his wider theory, just one more indication of the sort of thing going on seemingly everywhere else in the solar system, but it seems to me far more than that. As I say, it seems to me almost dispositive—at the very least, it seems the sort of account (so much more tangible than the existence of some putative Planet X that nobody has yet ever been able to see) that forces the onus back on those who would so cavalierly dismiss Murch's speculations.

For what is their counterproposal?

*

THE WEEKS PASS. I am always on the verge of thinking that I am almost finished with this narrative, but my inbox continues to swell, often with belated responses by Murch to earlier expressed criticisms. Thus, for example—partly in response to Smolin's surmise that if such wavelike pressure fluctuations actually existed, surely we would by now have had evidence of them in terms of observed anomalies in space probe trajectories and the like—Murch one day passes along a piece from BLDGBLOG. Under the title "Fossils of Spacetime," the entry begins by citing a piece in the November 17, 2014, issue of *New Scientist*, headlined "Dark Matter Could Be Seen in GPS Time Glitches," noting how teams of scientists led by Andrei Derevianko at the University of Nevada, Reno, and Maxim Pospelov of (Smolin's own) Perimeter Institute in Waterloo, Canada, have been trying to account for patterns of minuscule errors in the time readings of the earth-girdling ring of hyperaccurate satellite clocks, fifty thousand miles in diameter, that manage the world's GPS system. "Derevianko is already mining 15 years' worth of GPS timing data," convinced that the kinks in the data might reflect "kinks or cracks in the quantum fields that permeate the universe" and, to be specific, evidence of dark matter's otherwise notoriously elusive "fingerprints." Granted, this is not in any way what Murch is proposing, but it does traffic in very much the same manner of investigation. (BLDGBLOG's ever-mischievous curator, Geoff Manaugh, for his own part, goes on to note how "If you take all of this to its logical conclusion, you could argue that, hidden in the tiniest spatial glitches of the built environment, there is evidence not only of space weather but even potentially of the solar system's passage through 'kinks' and other 'topological defects' of dark matter, brief stutters of the universe now fossilized in the steel and concrete of super-projects like bridges and dams.")

At one point Alan Duffy wrote that he wasn't sure if Murch actually believed in the possible physical reality of his space-time ripples, continuing, though, that

> as a dark matter expert I can assure you that there are no "shells" or ripples in the dark matter distribution around the solar system. The Solar System is flying through an invisible cloud of dark matter and, just as rain appears to race into your front window screen as you drive through it in a car, so too is there a dark matter "wind" rushing... at tens of thousands of miles per hour past us as the solar system hurtles around the galaxy.

To which Murch now responded by sending along another *New Scientist* piece, this time from the August 5, 2015, issue, with the headline "Earth May Have a Hairy Mane of Dark Matter Flowing around It," complete with a highly suggestive speculative illustration: "Rather than being distributed smoothly throughout galaxies," the article explained, "simulations show that dark matter particles should clump into

A hairy mane of dark matter

features like halos, discs and streams," and goes on to note how

> Gary Prezeau at the Jet Propulsion Laboratory, California, wondered what would happen if a stream of dark matter pierced a planet like Earth. He calculated that the planet's gravity would bend the particles' trajectories and focus them to a point. This "lensing" effect would concentrate dark matter along an axis passing through Earth's core, reaching densities about a billion times more than average at the focal point.

Again, by no means exactly what Murch was not-quite-hypothesizing, but not without its pertinence.

Similarly, Murch sent along a recent piece from the September 22, 2015, issue of *Quanta Magazine*, a profile (by Natalie Wolchover) of physical theorist Nima Arkani-Hamed (of Princeton's Institute for Advanced Study), one of the principal protagonists in the *Particle Fever* documentary Murch edited a few years back. At one point, the piece describes how Arkani-Hamed and his student Jaroslav Trnka had

> uncovered a multifaceted geometric object whose volume encodes the outcomes of particle collisions—beastly numbers to calculate with traditional methods. The discovery suggested that the usual picture of particles interacting in space and time is obscuring something far simpler: the timeless logic of intersecting lines and planes. Although the "amplituhedron" [as Arkani-Hamed and Trnka dubbed their object] initially described a simplified

version of particle physics, researchers are now working to extend its geometry to describe more realistic particle interactions and forces, including gravity.

This led to the passage Murch was particularly flagging for me:

> Arkani-Hamed's ultimate goal is to describe the entire cosmological history of the universe as a mathematical object. In unpublished work, he has begun finding patterns in cosmological correlations—the likelihood, for instance, that if two red stars lie 20 kiloparsecs apart, a blue star lies 50 kiloparsecs away from them both. These statistical patterns encode the history of the cosmos, like dinosaur bones buried in the sand. And as with particle collisions, he has found that these patterns can be represented as geometric volumes. Ultimately, he said, anywhere from 10 to 500 years from now, the amplituhedron and these cosmological patterns will merge and become part of a single, spectacular mathematical structure that describes the entire past, present and future of everything "in some timeless, autonomous way."

Obviously Arkani-Hamed's project is far more ambitious than Murch's, but again, not without its pertinence as a mode of approach.

Other times, Murch would chime in with freshly unearthed dinosaur bones from his own trove of earlier writings. Still smarting perhaps from the accusation that he sometimes "cherry-picked" his data, he e-mailed me:

Trolling through my Bodean backlogs, I dug this up.

It is a spreadsheet I put together several years ago of the first 400 numbered asteroids, finding their common "orbital center"—the weighted average of their different semi-major axes.

The closest-in of all these is 2.175 AU, the most distant is 4.278 AU.

If an outer asteroid is four times the volume of an inner one, it is counted as if it is four asteroids all orbiting at the same AU distance. This 4:1 ratio would drag the average center outward. And vice versa of course. Ceres, the biggest asteroid, is one-and-a-half-million times the volume of Agathe, the smallest on this list. So you can see what kind of a tug of war this is.

The four columns are: Name & Number • Diameter in km • Volume in cubic km • Semi-major axis in AU.

The accompanying spreadsheet went on for eight pages.

The result [Murch concluded] is 2.80243.
Bode would predict 2.8.
So it is accurate to 100.09%

It turned out there was a reason he had been rummaging about in his back files. He had recently taken to wondering whether the trough between peaks to either side within which the asteroid belt was couched might not itself evince subtroughs, mini-valleys, as it were, along its slopes, at the "halfway" and "the halfway-from-the-halfway" points—of the sort into which Neptune and several of the renegade moons in the wider system seemed to have settled. He suspected that the majority of the hundreds of asteroids—and he now intended to dig deeper into the literature, so as to develop figures for the next several hundred asteroids as well—would settle pretty close to the bottom of the valley, but that there might well be an identifiable spike in the number of asteroids stuck in the two renegade mini-valleys. He sent along an illustrative diagram by way of clarification:

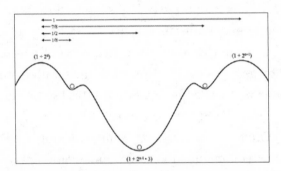

The mini-valley on the left is the point halfway between the two troughs to either side, which is to say those of Mars and Ceres, and the minivalley to the right constitutes halfway between the Ceres trough and the next halfway-between-troughs mini-valley that would occur an eighth of the way down the next wave over (the one with Jupiter lodged at its base). (Murch noted how in terms of musical notes, if the two barren peaks were denominated C and C an octave higher, the intervening notes would be D, G, and B-flat.) Just a hunch, he suggested, and one that might take him the next several months of spare time to nail down, but one that could potentially prove quite a fertile one, in terms of further dispositive proof.

(Around the same time, on an episode of NPR's *Radiolab*, I heard an account of efforts to register incredibly elusive dark matter by way of the exquisite monitoring of extraordinarily stilled pools of xenon buried deep, deep underground in an abandoned Colorado mine shaft, the only sort of place where sufficient silence, as it were, could be achieved to register evidence of such infinitesimally tiny effects. As part of the piece, the experimental physicist Rick Gaitskell was quoted as explaining how "Dark matter is the dominant component in the universe. The stuff you and I are made of, all these conventional protons and such, are the flotsam and jetsam of the material world, cast on a sea of dark matter." What, I found myself thinking, might it be like if instead of going to such remarkable lengths to achieve xenon-silence way, way deep underground, scientists were instead to take Gaitskell's sea metaphor seriously, as Murch was in fact already doing, and apply it to the sorts of asteroid belt inventories that Murch would now himself be attempting?)

Nor were such spreadsheets the only sorts of fossils Murch was meanwhile uncovering. Another time, he wrote:

Just ran across this for sale on eBay, one of those 35¢ Bantams, but with the cover done by my father, sometime in the 1950's—trying to micrometer the dimensions of space, planets and eccentric tilted orbits.

How we come to know what we think we know, indeed…
Gravitationally yours,
W. x

Three illustrations by Walter's father, W. T. Murch

A few days after this he sent me another image of his father's, a woodcut illustration from 1931 for Stuart Chase's *Men and Machines,* noting how "It looks like the apple does not fall far from the tree."

And a few days after that, Murched e-mailed that he'd woken up in the middle of the night

> realizing that there was another image we should include in your book, for reasons that will become obvious: In 1930, my father did the cover and illustrations for *The Mysterious Universe* by James Jeans FRS [Fellow of the Royal Society], a seminal book, summarizing the state of physics at that time and the impact that recent developments of quantum mechanics had on our view of the Universe. It was part of the intellectual furniture of my life growing up, and now that I think about it, the cover illustration obviously resonates with the imagery of Bode. This book, incidentally, was the source of that oft-quoted line

of Jeans: "The universe begins to look more like a great thought than a great machine."

Others of Murch's notes didn't so much raise up buried fossils as sound out tonal echoes. At another point, he wrote how:

> I was just reading a chapter last night in Koestler's *Ghost in the Machine* which I will scan and send, about neoteny and the necessity (in evolution, science, and art) for regression in order to then change angle of approach and move ahead when there is a blockage of some kind—what he describes as *reculer pour mieux sauter.*

The chapter in question (7), headed by an uncannily resonant epigram from Christina Rossetti ("Who has seen the wind? Neither you nor I. / But when the trees bow down their heads / The wind is passing by.") goes on to evoke and explore an evolutionary phenomenon known as paedomorphosis, which can occur when branching specialization reaches a dead end, and in order to move forward, progress first reverts back a few steps before branching out in a novel direction. "The crucial point here," Koestler points out, "is the appearance of some useful evolutionary novelty in the *larval or embryonic* stage of the ancestor, a novelty that may disappear before the adult reaches the adult stage, but which reappears and is preserved in the *adult stage of the descendent.*" Thus, for example, fishes descend from sea urchins, but not from adult sea urchins; rather, from their larval free-floating stage. Some develop gills and branch off into phyla of their

Koestler

own. Likewise, chimpanzees and humans, which share many, many embryonic characteristics not found in other primates. Koestler concludes by arguing that this distinctive diagonally back-and-forth, zigzag progression "could also represent a fundamental aspect of the history of ideas." Regarding all of which, Murch now glossed:

> And my own Bode project is like that, certainly: going back two and a half centuries for Bode, and then back to Pythagoras for celestial music—something that is unredeemably alien to the present scientific world view. As I have said (frequently), going to the Music of the Spheres was not my intention—it arose unbidden out of the material itself after a couple of months of work.
>
> And of course by regressing through these caverns we may discover something along the way—as Kepler did when trying to prove the truth of Platonic solids hovering in the interstices of space, and coming up with his three laws (not numbered nor considered especially important by him).
>
> Also I think it interesting: why did I, in particular, fall into this rabbit hole (which led into Chauvet-type caverns)? What combinations of mindsets do I have that made this likely, and why have I persisted for twenty years (pace

Smolin) pursuing this? What would Smolin have said to Kepler's twenty-year pursuit of Platonic Solids? What in fact did Galileo say to Kepler? (Galileo never accepted the idea of elliptical orbits, and believed Kepler's idea of the Moon somehow influencing the tides on Earth was "a joke.")

As if to answer that last brace of questions, a few days later Murch passed along some passages from a piece he had just been reading on the *New York Review of Books* blog (September 21, 2015), a translation of Roberto Calasso's observations on the occasion of Oliver Sacks's eightieth birthday a few years earlier, across which Calasso had quoted Sacks, variously, as writing

> I used to delight in the natural history journals of the nineteenth century, all of them blends of the personal and the scientific—especially Wallace's *The Malay Archipelago*, Bates's *Naturalist on the River Amazon*, and Spruce's *Notes of a Botanist* and the work which inspired them all (and Darwin too), Humboldt's *Personal Narrative*...
>
> They were all, in a sense, amateurs—self-educated, self-motivated, not part of an institution—and they lived, it sometimes seemed to me, in a halcyon world, a sort of Eden, not yet turbulent and troubled by the almost murderous rivalries which were soon to mark an increasingly professionalized world (the sort of rivalries so vividly portrayed in H. G. Wells's story "The Moth").

Sacks

This sweet, unspoiled, preprofessional atmosphere, ruled by a sense of adventure and wonder rather than by egotism and a lust for priority and fame, still survives here and there, it seems to me, in certain natural history societies, and amateur societies of astronomers and archaeologists, whose quiet yet essential existences are virtually unknown to the public. It was the sense of such an atmosphere that drew me to the American Fern Society in the first place, and that incited me to go with them on their fern-tour to Oaxaca early in 2000.

To which, Calasso now added, "Every time I read one of Oliver's writings, I find myself in that 'halcyon world' of science." And Murch in turn glossed:

That's a good summary of much of the appeal of Titius-Bode for me, independent of whether it is ever accepted by the institutional gatekeepers and Powers That Be. Indeed, sometimes independent of whether it is even "true" or not, by whatever standards "we" use to judge these things. That spreadsheet I sent you recently, on the common weighted center of the orbits of the first 400 asteroids, and its remarkable and beautiful (to me) result is often reward enough. This research has certainly provided me with a fascinating education into historical astronomy from Ptolemy through Copernicus, Kepler, and Titius-Bode, and the millennial progress of the human analytical mind. And, no matter what else it may yet prove to be, the math/music itself is a "halcyon world," a soothing balm for the sometimes rough abrasions of the film business.

CODA

MONTHS PASSED, AND suddenly the press began perco-
lating with word from a team of astronomers that they had
discovered evidence of a fresh and quite mammoth (Uranus-
or Neptune-scale) exoplanet orbiting way out beyond Pluto
after all, and it was quickly dubbed Planet 9 (Pluto itself having been stripped of that designation several years earlier).
I asked Murch by phone whether the discovery undermined
his Kuiper Cliff conjectures, to which he replied, "Not necessarily," continuing, "For one thing they still haven't sighted
the actual planet. Their claims are based instead on observed
perturbations in the orbits of several Kuiper Belt objects,
variations which they surmise could only be accounted for by
the gravitational pull of such a planet, which indeed everyone is now busy looking for. But such a planet, by their own
calculations, would be way, *way* out there. Keep in mind that
the sharp outer edge of the Kuiper Belt is at something like
50 AUs, whereas the orbit of this hypothetical planet, an
extremely elliptical one at that, is calculated to course somewhere between 1200 and 200 AUs, once every ten to twenty
thousand years—one has a hard time seeing how the resultant highly variable gravitational pressures could account for
the sharp suddenness and regularity of that 50 AU Kuiper
Cliff falloff. Not that there couldn't be some *other* planet out

there doing that, it's just that they still haven't found any other evidence for that one. As for Planet 9, if it ever is actually found, it will be fascinating to see what its semi-major axis proves to be: they're saying it might be around 700 AUs, and there *is* a Titius-Bode trough at 614.8 AUs. So, we'll just have to wait and see."

A few weeks after the Planet 9 report, on February 11, 2016, scientists affiliated with the billion-dollar LIGO Lab (the Laser Interferometer Gravitational-Wave Observatory— actually two such extremely high-tech installations working in millisecond concert, one in Livingston, Louisiana, and the other in Hanford, Washington) made the epochal announcement that, just over a hundred years on, they had been able to confirm a central prediction of Einstein's general theory of relativity, by registering the passage of an actual gravitational wave traveling at the speed of light and generated from the collapse of two black holes into each other, 1.3 billion light years from Earth (and hence 1.3 billion years ago)—a wave that had actually swept past earth, registering its infinitesimally tiny ping a few months earlier, on September 14, 2015. (Thousands of scientists had been busily engaged in confirming and double- and triple-checking the astonishingly precise reading ever since.)

Murch, by e-mail, was just as impressed with the announcement as everyone else, though he reiterated that it had little immediate bearing on the kind of thing he had been postulating in his Titius-Bode speculations.

When I speak of undulations in the fabric of space-time radiating out from large solar or planetary gravitational centers, remember, I am speaking in a Vaihingerian As-If mode, a metaphoric image which gives a feeling for what I am talking about. A gravity wave of the sort that may just have been identified at LIGO is different: a propagating pulse of energy which travels at the speed of light. What I am alluding to is, rather, a series of *standing* waves in the vicinity of massive objects such as our Sun and its planets and perhaps other stars and their planets in the Galaxy. Because it is *as if* there is an underlying modulation—a rippled structure—to the space-time fabric of the solar system which keeps the planets in their orbits.

What the real reality of this is, again, I don't know. Could it be fluctuations in the density of hypothetical dark matter? Time? Green cheese? Or something else?

As Bertrand Russell once wrote: "How things seem to seem is not enough. We must somehow discover how things *really* seem!"

Another of Walter's fortune cookies, I assumed.

"Speculating about the causes of this is fascinating, of course, but has to be treated carefully," he reiterated.

One needs to impose a kind of magnetic bottle around such discussions, to create a safe zone where things are prevented from being interpreted literally. These undulations *in something* are metaphoric images, useful to get across the general idea. What the actual cause is, I don't know and am not competent to get into specifics. But my basic argument remains: there is a high degree of regularity in the spacing of the planets and moons of the solar system, which follows a Bodean pattern (with Murchean modifications), and it should be investigated further to discover the cause. We can speculate and create metaphorical images that might be helpful. But that is all they are at this stage: metaphorical images.

Having provided that disclaimer, it's worth noting how the Quantum nature of reality was first proposed by Max Planck, in 1900, as a *metaphorical* solution to the Violet Catastrophe [long story: Google it] but even Planck himself was dubious about it. The physics community only woke up to its *reality* when Einstein explained the photoelectric effect using it. Until then, the quantum had seemed to be a mathematically useful fiction, though without any physical reality. As the decades passed, however, this statistical *as-if* solidified (if that is the right word) into quantum mechanics, upon which our entire present-day vision of the micro-universe is based, driving our iPhones and all the other devices.

And, who knows, someday something similar may yet be seen to be going on with these seemingly seeming Titius-Bode undulations.

Who knows, indeed.

"Perseverence furthers," as Murch himself has recently taken to signing off his various missives, quoting his old friend (as he puts it), "Mr. I. Ching."

ACKNOWLEDGMENTS AND SOURCES

FOR STARTERS, OF course, my deepest gratitude with regards to this project goes to its principal subject, the remarkably generous, capacious, and patiently forbearing Mr. Murch himself. Time will tell whether he turns out to have been right with regard to all the speculations unspooled across this book (it may yet have proven useful, if nothing else, to have lain down a marker). What has been most striking to me throughout our work together, however, has been Murch's own equanimity in this regard, and notwithstanding the evident enthusiasm and intensity with which he has pursued this particular passion, his openness to the possibility that he may indeed be wrong ("Apophenia!") and his acknowledgment that, no matter what, the intimations and implications of what he is convinced he is seeing remain hazy and incomplete at best. That modesty, that unfailing sense of proportion, and the constant good humor in which they are cast have characterized all of my interactions with Murch over the coming-on two decades of our deepening friendship (for this was not our first outing, see below)—and for all of that, again, I thank him. And I thank his splendid consort, Aggie, as well (talk about unending forbearance!).

And then of course I owe a debt of gratitude to all of those experts who agreed to talk with me, and in so doing to school me in the science, history, and sociology behind many of the issues raised in this book: Charles Falco, Richard Greenberg, Lawrence Krauss, Robert Speare, Alan Duffy, Stuart Firestein, George Blumenthal, the late Dorothy Nelkin, and especially Margaret Wertheim and Lee Smolin. (Among mere civilians, I'd also include Michael Benson, Beth Jacobs, Lena Herzog, Natalie de Souza, Robert Krulwich, Hugh and Nat Osborn, David Gersten, Gerri Davis, and my "chief of staff" Laura deBuys.) Collectively they all helped me to avoid a whole slew of infelicities and mischaracterizations; to the extent that any others have persisted into this final text, those are entirely my own stubborn fault, and not for lack of their wise attempted interventions.

On the writerly side of things, once again I am in serious debt to my marvelously supportive agent Chris Calhoun, and to ongoing editor friends Luke Mitchell and Alex Star (who was kindly forbearing, that word again, of my need to wrap up this project before proceeding on to the next, which I have promised to him), and especially to the entire team at Bloomsbury, starting with my brimmingly attentive editor George Gibson (with whom I've been looking forward to working for years), his assistant Callie Garnett, the art director Patti Ratchford, and the designer Sara Stemen.

And finally, as ever, I appreciate beyond words the support and the ever put-upon forbearance (one last time) of my lovely bride, Joasia, and our vivid daughter, Sara, and her dear friend (effectively, our new daughter), Niwa, to whom this book is happily dedicated.

*

AS FOR SOURCES, they tend to range themselves into various categories, as follows:

BY OR ABOUT MURCH

The iteration of Walter Murch's Music of the Spheres lecture specifically referenced in this book—"New Evidence Confirms 18th Century Conjecture on Orbital Harmonies"—was given at the New School at Commonweal in Bolinas, California, in April 2015, and introduced by the school's founding director, Michael Lerner. A video of the event can be accessed here: https://vimeo.com/125379787. (My deep appreciation to everyone at Commonweal for their enthusiastic hospitality and cooperation.)

I also drew on several texts by Murch himself, notably including his unpublished treatise on Titius-Bode, "Skybound," but also:

> *The Bird that Swallowed Its Cage: The Selected Writings of Curzio Malaparte.* Translated and with an introduction by Walter Murch. Afterword by Lawrence Weschler. Berkeley, CA: Counterpoint, 2013.

> *In the Blink of an Eye: A Perspective on Film Editing.* 2nd ed. Los Angeles, CA: Silman-James Press, 1995/2001.

> "Oblivion or Immortality," Murch's essay on the uncanny convergence of Kepler and Rosenkranz, which ran in *Brick* 88, Winter 2012.

> "Painting the Air: Reminiscences of Life with Walter Tandy Murch," which will be included in *Walter Tandy Murch: Paintings and Drawings, 1925–1967*, due out from Artists Book Foundation, New York, in 2017.

> "The Three Fathers of Cinema: Beethoven, Flaubert, & Edison," video of a twenty-minute presentation by Murch at a Spanish cinema conference: https://vimeo.com/11117217.

> Murch's extended residency on Transom.org in the spring of 2005, which is still archived at http://transom.org/2005/walter-murch/ and includes two long manifestos by Murch,

"Womb Tone" (drawing on his wife Aggie's radio piece "A Mother's Symphony") and "Dense Clarity/Clear Density," the latter featuring a detailed discussion of the Rule of Two and a Half as it relates to robotic footfalls in *THX 1138* and the Valkyrie scene in *Apocalypse Now*, with examples drawn from the editing of both films; and extended responses to myriad subsequent correspondents.

In addition, valuable works by others *about* (and to a significant extent *with*) Murch include:

Michael Ondaatje. *The Conversations: Walter Murch and the Art of Editing Film.* New York: Knopf, 2002.

Charles Koppelman. *Behind the Seen: How Walter Murch Edited Cold Mountain Using Apple's Final Cut Pro and What This Means for Cinema.* San Francisco: New Riders Press, 2004.

Geoff Manaugh's interview with Murch, "The Heliocentric Pantheon: An Interview with Walter Murch," which can be accessed at Manaugh's website: http://www.bldgblog.com/2007/04/the-heliocentric-pantheon-an-interview-with-walter-murch/

As well as my own: "Valkyrie over Iraq: Walter Murch, *Apocalypse Now, Jarhead,* and the Trouble with War Movies," originally published in *Harper's,* November 2005, and subsequently included in my collection *Uncanny Valley: Adventures in the Narrative.* Berkeley: Counterpoint, 2011.

Bill Morrison, the filmmaker, included a remarkable visualization of Murch's Bodean ideas about Jupiter and its moons in the middle movement, "Chorale," of his 2012 short film *Just Ancient Loops,* with music by Michael Harrison performed by the cellist Maya Beiser, and CGI visualizations rendered with the collaboration of the Advanced Visualization Laboratory of the National Center for Supercomputing Applications at the University of Illinois, Champagne-Urbana. "Chorale" can be accessed online at vimeo.com//178421654.

BOOKS AND ARTICLES REFERENCED OR
CONSULTED ON THE HISTORY OF ASTRONOMY
AND TITIUS-BODE IN PARTICULAR

Lincoln Barnett. *The Universe and Dr. Einstein*. New York: William Morrow, 1957.

Arthur Koestler. *The Sleepwalkers*. New York: Macmillan, 1959; Danube edition, 1968.

Michael Martin Nieto. *The Titius-Bode Law of Planetary Distances: Its History and Theory*. Oxford: Pergamon Press, 1972.

Edgar Allan Poe. *Eureka: A Prose Poem on the Material and Spiritual Universe*. Originally published by George Putnam, New York, 1848. A special edition for distribution at the *Eureka* exhibition at the Pace Gallery, NYC, 2015.

THE BOVAIRD/LINEWEAVER PAPERS ON TITIUS-BODE
IN EXOPLANET SYSTEMS AND VARIOUS RESPONSES

In 2013, associate professor Charles Lineweaver, working with his grad student Timothy Bovaird, both of the Australian National University, published their original paper, "Exoplanet Predictions Based on the Generalized Titius-Bode Relation," in *Monthly Notices of the Royal Astronomical Society* 435, no. 2, pp. 1126–38.

In 2014, Chelsea Huang and Gaspar Bakos (both out of Princeton) published a largely critical paper on the Lineweaver/Bovaird thesis, "Testing the Titius-Bode Law Predictions for Kepler Multiplanet Systems," in *Monthly Notices of the Royal Astronomical Society* 442, no. 1, pp. 674–81.

In 2015, Bovaird and Lineweaver, joined by Steffen Kjær Jacobsen (of the Niels Bohr Institute in Copenhagen) followed up with a paper responding to the Huang-Bakos critique and expanding on some of their own original notions: "Using the Inclinations of Kepler Systems to Prioritize New Titius-Bode-based Exoplanet Predictions," also in *Monthly Notices of the Royal Astronomical Society* 448, no. 4, pp. 3608–27.

A wealth of responses and counterresponses to these papers quickly grew online, but a good introductory overview was provided by the science writer Paul Gilster on his Centauri Dreams weblog in "Can We Find Exoplanets Using the Titius-Bode Relation?" http://www.centauri-dreams.org/?p=32757.

ON THE PHILOSOPHY AND SOCIOLOGY OF SCIENCE

Hasok Chang. *Is Water H₂O? Evidence, Realism and Pluralism.* Heidelberg: Springer Publishing, 2012.

Stuart Firestein. *Ignorance: How It Drives Science.* Oxford/New York: Oxford University Press, 2012.

David Graeber. *The Utopia of Rules: On Technology, Stupidity, and the Secret Joys of Bureaucracy.* Brooklyn: Melville House, 2015.

Lee Smolin. *The Trouble with Physics: The Rise of String Theory, the Fall of a Science, and What Comes Next.* New York: Mariner Books, 2006.

P. Kyle Stanford. *Exceeding Our Grasp: Science, History, and the Problem of Unconceived Alternatives.* Oxford/New York: Oxford University Press, 2006.

Hans Vaihinger. *The Philosophy of "As If": A System of the Theoretical, Practical and Religious Fictions of Mankind.* London: Routledge and Kegan Paul LTD, 1924/1949.

Margaret Wertheim. *Physics on the Fringe: Smoke Rings, Circlons, and Alternative Theories of Everything.* New York: Walker & Company, 2011.

Susan Wise Bauer. *The Story of Western Science from the Writings of Aristotle to the Big Bang.* W. W. Norton & Company, New York, 2015.

WALTER MURCH'S OWN LIST

As we were going to press, I asked Walter Murch for a list of readings that had been important to him during the years of his research, and, in addition to the Koestler (*The Sleepwalkers*), the Nieto (*The Titius-Bode Law of Planetary Distances*), and the Poe (*Eureka*) cited above, he offered:

By way of general historical and scientific books

Daniel J. Boorstin. *The Discoverers: A History of Man's Search to Know His World and Himself.* New York: Random House, 1983 (which is where Walter says he first came upon the delicious Kepler salad/wife reference).

Kitty Ferguson. *Tycho & Kepler.* New York: Walker & Company, 2002.

Timothy Ferris. *Coming of Age in the Milky Way.* New York: William Morrow & Co., 1988.

Ivars Peterson. *Newton's Clock: Chaos and Order in the Solar System.* New York: W. H. Freeman & Co., 1995.

Specifically with regard to musical/cosmological aspects

Ross Duffin. *How Equal Temperament Ruined Harmony: (and Why You Should Care).* New York: W. W. Norton & Company; 2008.

Jamie James. *The Music of the Spheres: Music, Science and the Natural Order of the Universe.* Göttingen, Germany: Copernicus Publication, 1993.

James Jeans. *Science and Music.* Cambridge: Cambridge University Press, 1927.

Thomas Levenson. *Measure for Measure: A Musical History of Science.* New York: Touchstone, 1995.

The old ones

Nicolaus Copernicus. *On the Revolutions of the Heavenly Spheres.* Translated by Charles Wallis. Amherst, NY: Prometheus Books, 1995.

Nicolaus Copernicus. *Three Treatises.* Translated by Edward Rosen. Mineola, NY: Dover Publications, 2004.

Johannes Kepler. *The Epitome of Copernican Astronomy and Harmonies of the World.* Translated by Charles Wallis. Amherst, NY: Prometheus Books, 1995.

Johannes Kepler. *Mysterium Cosmographicum* (Murch couldn't find his translated copy of the book itself but noted that the original was published in Tübingen in 1596).

IMAGE CREDITS

Every effort has been made to contact all copyright holders. If notified, the publisher will be pleased to rectify any errors or omissions at the earliest opportunity.

Page viii: Murch astride the Greenwich Meridian © Matthew Robbins
Pages 3, 4, 5, 36, 38, 39, 40, 41, 44, 45, 50, 51, 52, 57, 58, 61, 62, 64, 134: Slides from Walter Murch Keynote address
Pages 6, 7, 8, 11, 22, 26, 28, 34, 53, 54, 72, 120, 140, 142: Courtesy of Walter Murch
Page 10: John Cage © James Klosty
Page 12: Zoetrope gang © Francis Ford Coppola Presents
Page 14: Gerri Davis
Page 18: © Courtesy of Lucasfilm Ltd. LCC.
Page 20: *Apocalypse Now* © Francis Ford Coppola Presents
Page 35: *Rosencrantz and Guildenstern are Dead* © Grove Atlantic
Page 145: Oliver Sacks © Roberto Calasso

INDEX

NOTE ON THE AUTHOR

Lawrence Weschler is a critic, journalist, and author who was a staff writer at the *New Yorker* for more than twenty years. His work includes two earlier books in the current series: *Mr. Wilson's Cabinet of Wonder,* for which he was short-listed for the Pulitzer Prize and the National Book Critics Circle Award, and *Boggs: A Comedy of Values.* Others of his books include *Everything That Rises,* which received the 2007 National Book Critics Circle Award for criticism, *Seeing Is Forgetting the Name of the Thing One Sees,* and *Vermeer in Bosnia.* He has written for *Vanity Fair,* the *New York Times,* the *Nation, Salon, Truthdig,* and *Harper's,* among others.